SpringerBriefs in Applied Sciences and Technology

Thermal Engineering and Applied Sciences

Series Editor
Francis A. Kulacki, University of Minnesota, USA

For further volumes:
http://www.springer.com/series/10305

Dennis A. Siginer

Stability of Non-Linear Constitutive Formulations for Viscoelastic Fluids

Springer

Dennis A. Siginer
Centro de Investigación en Creatividad
y Educación Superior
Universidad de Santiago de Chile
Santiago, Chile

ISSN 2191-530X ISSN 2191-5318 (electronic)
ISBN 978-3-319-02416-5 ISBN 978-3-319-02417-2 (eBook)
DOI 10.1007/978-3-319-02417-2
Springer Cham Heidelberg New York Dordrecht London

Library of Congress Control Number: 2013953335

Printed on acid-free paper

Springer is part of Springer Science+Business Media (www.springer.com)

Preface

This monograph together with its complimentary volume, D. A. Siginer, *Dynamics of Tube Flow of Viscoelastic Fluids*, Springer, New York, 2014 in this series is an attempt to give an overall comprehensive view of a complex field, which has developed rapidly in the second half of the twentieth century, still far from being settled on firm grounds, that of non-linear constitutive equation formulations for viscoelastic fluid media and their impact on the dynamics of viscoelastic fluid flow in tubes. This monograph covers the development of constitutive equation formulations in their historical context together with the latest progress made, and *Dynamics of Tube Flow of Viscoelastic Fluids* covers the state-of-the-art knowledge in predicting the flow of viscoelastic fluids in tubes highlighting the historical as well as the most recent findings. Most if not all viscoelastic fluids in industrial manufacturing processes flow through tubes, which are not necessarily circular, at one time or another during the processing of the material. It is critically important that the flow of non-linear viscoelastic fluids in tubes can be predicted on a sound basis, thus the *raison d'être* of the volume on *Dynamics of Tube Flow of Viscoelastic Fluids*. As flow behavior predictions are directly related to the constitutive formulations used, *Dynamics of Tube Flow of Viscoelastic Fluids* relies heavily on this volume when viscoelastic fluids are the working fluids.

The science of rheology defined as the study of *the deformation and flow of matter* was virtually single handedly founded and the name invented by Professor Bingham of Lafayette College in the late 1920s. Rheology is a wide encompassing science which covers the study of the deformation and flow of diverse materials such as polymers, suspensions, asphalt, lubricants, paints, plastics, rubber, and biofluids, all of which display non-Newtonian behavior when subjected to external stimuli and as a result deform and flow in a manner not predictable by Newtonian mechanics.

The development of rheology, which had gotten to a slow start, took a boost during WWII as materials used in various applications, in flame throwers for instance, were found to be viscoelastic. As Truesdell and Noll famously wrote (Truesdell C, Noll W (1992) The non-linear field theories of mechanics, 2nd edn. Springer, Berlin) "By 1949 all work on the foundations of rheology done before 1945 had been rendered obsolete." In the years following WWII the emergence and rapid growth of the synthetic fiber and polymer processing industries, the appearance of liquid detergents, multigrade oils, non-drip paints, and contact adhesives, and developments in pharmaceutical and food industries and biotechnology spurred the development of rheology. All these examples clearly illustrate the relevance of rheological studies to life and industry. The reliance of all these fields on rheological studies is at the very basis of many if not all of the amazing developments and success stories ending up with many of the products used by the public at large in everyday life.

Non-Newtonian Fluid Mechanics, which is an integral part of rheology, really made big strides only after WWII, and has been developing at a rapid rate ever since. The development of reliable constitutive formulations to predict the behavior of flowing substances with non-linear strain–stress relationships is quite a difficult proposition by comparison with Newtonian fluid mechanics with linear stress–strain relationship. The latter does enjoy a head-start of two centuries tracing back its inception to Newton and luminaries like Euler and Bernoulli. With the former the non-linear structure does not allow the merging of the constitutive equations for the stress components with the linear momentum equation as it is the case with Newtonian fluids ending up with the Navier–Stokes equations. Thus the practitioner ends up with six additional scalar equations to be solved in three dimensions for the six independent components of the symmetric stress tensor. The difficulties in solving in tandem this set of non-linear equations, which may involve both inertial and constitutive non-linearities, cannot be underestimated. Perhaps equally importantly at this point in time in the development of the science we are not fortunate enough to have developed a single constitutive formulation, which may lend itself to most applications and yield reasonably accurate predictions together with the field balance equations. The field is littered with a plethora of equations, some of which may yield reasonable predictions in some flows and utterly unacceptable predictions in others. Thus we end up with classes of equations that would apply to classes of flows and fluids, an ad hoc concept at best that hopefully will give way 1 day to a universal equation, which may apply to all fluids in all motions. In addition the stability of these equations is a very important issue. Any given constitutive equation should be stable in the Hadamard and dissipative sense and should not violate the basic principles of Thermodynamics.

Efforts have not been spared to be thorough in the presentation with commentaries about the successes and failures of each theory and the reasons behind them. The link between different theories and the naturally unfolding succession of theories over time borne out of the necessity of better predictions as well as the challenges in the field at this time are given much emphasis at the expanse of a detailed in-depth development of various theories. This monograph provides a snapshot of a fast developing topic and a bridge connecting new research results with a timely and comprehensive literature review. For a detailed in-depth exposition of any one subject included in this book the reader is referred to the extensive reference list given at the end of each chapter. The responsibility for any mistakes and misquotes that may have crept up into the text in spite of extensive checking remains solely with the author.

Santiago, Chile Dennis A. Siginer

Contents

Chapter 1
Preamble

Abstract The nature of viscoelasticity, difficulties in proving the existence and uniqueness of the solution of the set of field equations defining a given viscoelastic fluid flow problem, which form the basis of numerical solutions, the high Weissenberg problem, limiting value of the Weissenberg number in computations, the role of singularities and mesh size in computational Rheology, Hadamard instability, and well-posed Cauchy problems are discussed in setting up the stage for constitutive formulations for non-linear viscoelastic fluids.

Keywords Linear and non-linear viscoelasticity • High Weissenberg number problem • Role of singularities • Computational Rheology • Hadamard instability • Well-posedness

This monograph presents an overall view of the theories and attendant methodologies developed independently of thermodynamic considerations as well as those set within a thermodynamic framework to derive rheological constitutive equations for non-linear viscoelastic fluids with due emphasis on the historical context. Developments in formulating Maxwell-like constitutive differential equations as well as single integral constitutive formulations are discussed in the light of Hadamard and dissipative type of instabilities they may be inherently vulnerable. The field is by no means mature, and there are many unresolved issues. For instance constitutive formulations (CE) and their stability are central to the prediction of secondary flows of viscoelastic fluids in tubes, a topic of substantial importance to process industries still unresolved from the analytical/computational predictions point of view. The progress made in non-linear CE formulations is summarized, and the perplexing issues still plaguing the field are reviewed.

Historical and recent developments concerning constitutive formulations, including the nature of viscoelasticity, linear versus non-linear viscoelasticity, constitutive formulations based on local action versus non-local action, linear and

D.A. Siginer, *Stability of Non-Linear Constitutive Formulations for Viscoelastic Fluids*, SpringerBriefs in Applied Sciences and Technology 14,
DOI 10.1007/978-3-319-02417-2_1,

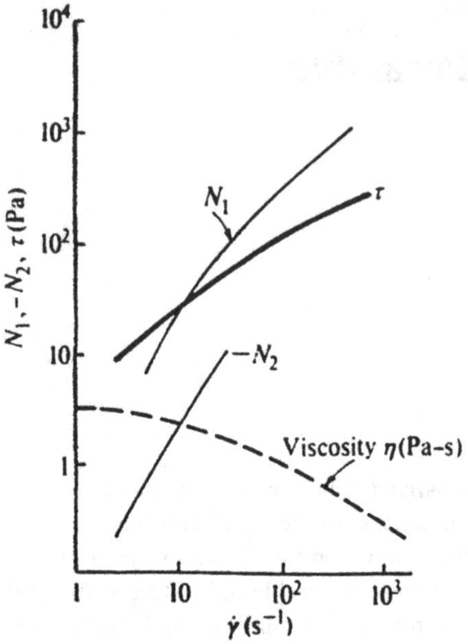

Fig. 1.1 The first and second normal stress differences N_1, N_2, the shear stress τ, and the viscosity η as function of shearing rate for a 6.8 % polyisobutylene in cetane solution at 24 °C (Reprinted from Tanner [1] with permission)

non-linear viscoelastic constitutive equations (CE), thermodynamically compatible non-linear viscoelastic CE formulations, and the stability of popular Maxwell-like differential and single integral type constitutive formulations are covered as well as their relationship to the observed physical instabilities. Hadamard and dissipative stable CE formulations firmly rooted in a thermodynamic basis capable of predicting if not quantitatively but at the very least qualitatively all the essential features of flowing viscoelastic fluid behavior such as shear-rate-dependent viscosity, at least uniaxial elongational behavior with elastic recoil, and second normal stresses is still an elusive goal. The complexity of the field equations at the macroscopic level given the non-linear nature of the CE and therefore the impossibility of combining the linear momentum balance with the CE and ending up with a universal field equation as in the case of Newtonian fluids makes analytical solutions intractable in almost all problems of industrial interest. As a consequence numerical solutions to non-Newtonian fluid flow problems gain another dimension in importance. However, computational difficulties even in attempts to qualitative predictions are still unresolved and get worse with increasing Weissenberg numbers *We* ultimately hitting a wall and breaking down. This is the so-called High Weissenberg Number Limit (HWNL) in numerical computations with non-linear viscoelastic fluids. Increased computer power together with finer mesh size is not the solution as the numerical algorithms blow up no matter how fine the mesh size may be; actually the blow-up gets much worse as the mesh size gets smaller. The non-linear viscoelastic fluids, whose behavior modeling under forcing is the main subject of this monograph, display the following general physical characteristics under shear, Figs. 1.1 and 1.2.

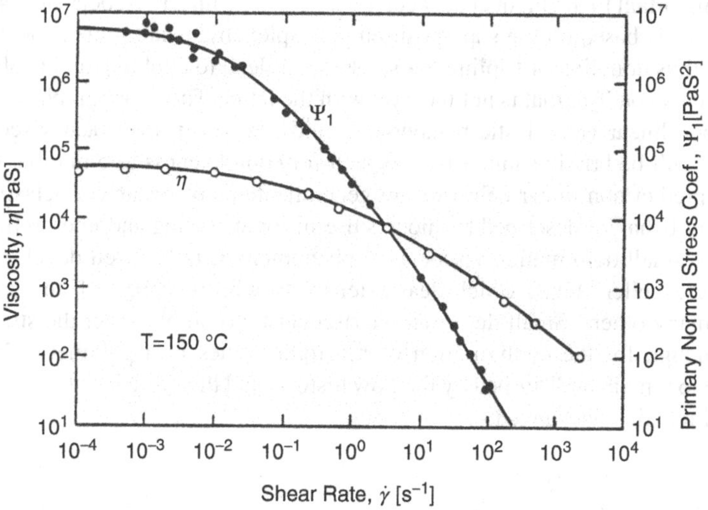

Fig. 1.2 Steady-state viscosity, and the first normal stress coefficient versus shear rate for polyethylene (Reprinted from Laun [2] with permission)

1.1 Viscoelasticity

All materials possess some degree of viscoelasticity depending on the time scale of the process. Which of the viscous and elastic properties will be dominant is defined by the natural time of the material and the time scale of the particular process the material is undergoing. Thus, if the process is relatively fast, that is it takes place in a time span smaller than the natural time of the material processed, the material will appear to be elastic rather than viscous and will be said to be elastic showing Hookean behavior. In the opposite situation, that is if the process time is larger than the natural time, the same material sample will be said to be viscous or shows Newtonian behavior. If the process takes place at intermediate time scales, a mixed response may be observed and the material will be labeled as showing *viscoelastic* behavior in these time scales. Thus, the classification is dependent on the interaction between the nature of the material and the nature of the observation. The distinction between "solid" and "fluid" is usually made on the basis of a subjective comparison of the relaxation time and the period of observation. A good example is glass, which is a supercooled liquid at room temperatures and flows very slowly noticeable only over centuries and regarded as a solid over shorter periods. If the relaxation time is so short that stress relaxation escapes observation or it is so long that no stress relaxation is observed during the period of the observation it may be concluded that the material is perfectly "elastic" or a "fluid" as the case may be. The material is *viscoelastic* when the relaxation time and the period of observation are of the same order of magnitude.

It is important to distinguish between linear and non-linear viscoelastic behavior. The former is based on the superposition principle: strains are proportional to the stresses, thus doubling or tripling the stresses will lead to doubling and tripling the strains and so on. But that is not the case with the latter. This monograph will focus on the non-linear viscoelastic behavior of fluids; however, the linear viscoelastic behavior will be briefly summarized as well for completeness as a number of CEs used to predict non-linear behavior are generalizations of linear viscoelastic CEs. The linear behavior described by models useful for analyzing and interpreting data related to small deformation viscoelastic phenomena is rather well developed and there are excellent texts, which deal extensively with it, Ferry [3], to name one among many others. Small deformation viscoelasticity implies that the strains are small enough for the configuration of macromolecules in a polymeric fluid for instance to remain undisturbed by the flow history, and therefore the structure of the material remains unchanged.

1.2 Computational Approach to Solutions

It is important to comment briefly on the increasing surge over the last 2 decades in numerical attempts in the literature to bring a solution to numerous non-Newtonian flow problems both of industrial and theoretical significance. The search for a numerical solution makes sense if it can be shown that the solution of a set of differential equations possesses the properties of existence and uniqueness. However, for many steady viscoelastic fluid flows, particularly for those of industrial interest, unlike the much better developed mathematical theory of the existence and uniqueness of steady Newtonian fluid flow problems, existence and uniqueness proofs are extremely difficult if not impossible to establish. The set of conservation equations for mass and momentum in the Newtonian case is augmented by the addition of the constitutive equations, six additional scalar equations if the extra-stress tensor is assumed to be symmetric, which do complicate the picture substantially. However, the need for solutions to guide, predict, and design industrial flows in particular, necessarily leads to the adoption of a pragmatic approach, to take a leap of faith and seek numerical solutions of Cauchy problems for which existence and uniqueness results have not been established in the hopes that existence and uniqueness results will be established sometime in the future. One major difficulty in proceeding with existence and uniqueness of viscoelastic fluid flow problems is that perturbations pivoted about Navier–Stokes equations do not work because the field equations governing the flow of viscoelastic fluids are singular perturbations of the Navier–Stokes equations with higher order derivatives in the equations. Another major difficulty stems from the memory of viscoelastic fluids which do remember their past deformation history, the more recent the deformation the stronger its influence on the present state of the deformation. Therefore, in any investigation of existence and uniqueness of the flow in a particular domain information is required at the inflow boundaries concerning the deformation history of the particles entering the domain. There are a few results in the literature for steady flows of the upper-convected Maxwell (UCM) fluids and the matter rests there for the time being.

In spite of the lack of existence and uniqueness results attempts borne essentially out of necessity are being made at an increasing rate to successfully simulate viscoelastic flows numerically for all Weissenberg *We* and Deborah *De* numbers. It should be recalled that most industrial processes operate at high Deborah numbers of the order of several hundreds. Among the plethora of problems which stand in the way for successful simulations the outstanding issue is the "High Weissenberg number limit" (HWNL), which refers to the limiting value of the Weissenberg number beyond which numerical solutions fail. Steady viscoelastic fluid flows are governed by mixed-type elliptic–hyperbolic partial differential field equations (PDFE), for example, the system of PDFEs governing the unsteady flow of the Oldroyd-B fluids is neither strictly elliptic nor strictly hyperbolic. The presence of the hyperbolic component of the PDFE complicates the numerical solution to a great extent. This is unlike the PDFE for inelastic non-Newtonian fluids for which case the PDFEs are elliptic. The discretization error is more of a serious issue with hyperbolic type PDFEs because any error will propagate along the characteristics, but that is not the case with elliptic problems as there are no real characteristics. In the early developments, attempts were made to discretize the PDFEs with methods appropriate for elliptic equations such as the Galerkin method. We know now (thanks to Joseph et al. [4], Joseph and Saut [5], and Joseph [6]) that the standard Galerkin method is not appropriate for discretizing the viscoelastic PDFEs. As a result, the limiting value of the Weissenberg number has been pushed back to higher values. The contemporary research efforts in pushing back further HWNL revolve around adequately resolving stress boundary layers and flow structure around singularities in viscoelastic fluid flow, refer Owens and Phillips [7]. The presence of singularities is especially significant in flow domains with non-smooth boundaries in degrading the accuracy of numerical solutions. At the boundary points where the singularity exists, the solution or any of its derivatives will fail to be integrable, which triggers the degradation of the numerical solution through errors introduced in the numerical approximation in the vicinity of the singularity.

The mathematical basis for conventional discretization methods originally developed for solving elliptic PDEs, such as the finite difference and finite element methods, is firmly established. Even though viscoelastic problems are governed by mixed-type PDFEs and the analysis of a viscoelastic singularity is extremely complicated, it is instructive to look at the influence of the singularities on the solution of elliptic problems where theoretical error estimates are available unlike the viscoelastic case. Consider the Poisson problem in a domain Ω with boundary $\partial\Omega$,

$$\nabla^2\varphi = -f, \quad \varphi(x)\in\Omega, \quad \varphi(x)_{\partial\Omega} = 0, \quad x\in\Omega$$

If $\partial\Omega$ is smooth, the following holds $\varphi \in H^{s+2}(\Omega)$ if $f \in H^s(\Omega)$, which implies that φ is smooth if f is smooth. However, if $\partial\Omega$ has corners, the solution φ or its derivatives will have singularities at the corners even if f is smooth. In elliptic problems for some derivative of the solution, the error would behave like $O(h^{-s})$ for some $s > 0$ where h is a characteristic length associated with a finite element. It can be shown for instance that in the second-order central difference discretization in finite difference representation of the Laplacian near a corner with the angle $3\pi/4$ the error behaves like $O(h^{-4/3})$. Consequently, as $h \rightarrow 0$ the error becomes unbounded. Clearly mesh refinement is not the solution to resolving this singular behavior. If the local behavior of the solution in the vicinity of the singularity is known in elliptic problems, then there are remedies to treat the singularity and improve the accuracy of the finite difference, finite element, and spectral element solutions. In finite difference methods, the singularity may be subtracted out of the problem to obtain a nonsingular problem; in finite element approach, the basis of trial functions may be supplemented by appropriate singular functions, refer Fix et al. [8], and in spectral element methods, the singularity within a single element may be isolated and an expansion in terms of known singular functions may be used, refer Owens and Phillips [7].

It can be shown that the asymptotic behavior of the solution of the Navier–Stokes equations is Stokes like near the singularity, any singularity. The Newtonian problem for the plane steady flow near a corner formed by two inclined planes has been extensively investigated by Dean and Montagnon [9] who showed that a first approximation to the flow of a viscous liquid in the neighborhood of a boundary with a corner may be obtained by solving the linearized equations of momentum neglecting the inertia terms, and that the stream function ψ_m near enough to the corner is of the form $r^m f_m(\theta)$. Corners pointing into the flow (aperture angle α larger than 180°) generate singular stresses at the corner, which behave like r^{m-2}. The values of the exponent m depend on and vary with the aperture angle α. The relationship between m and α is governed by the characteristic equation $\sin \alpha(m - 1) = \pm (m - 1) \sin \alpha$, plus and minus signs on the RHS corresponding to antisymmetric and symmetric flow patterns about the symmetry line $\alpha/2$. The question of separation angle arises in the same context. Dean and Montagnon [9] show that when $\alpha < 146.3°$, vortices appear in the corner and the exponent m is complex. In follow-up work, Michael [10] has shown that separation occurs at $\alpha = \pi$. However, neither of these results is always confirmed by experiments and computations. Tanner [1] suggests that Newtonian fluid physics may be inadequate near the singularity and that further research is called for on this issue. Newtonian behavior is important because for viscoelastic models with a solvent viscosity near corner fluid behavior is more often than not dominated by the solvent viscosity as pointed out by Pearson and Richardson [11].

The form of the singularity in the case of generalized Newtonian fluids without memory, the so-called inelastic fluids, power law fluids with viscosity variation of the type $\eta = (\dot{\gamma})^{n-1}$, has been studied by Henricksen and Hassager [12] who determined that the velocity behaves like r^{α} with $\alpha = f(n)$ near the corner. An example for the range of shear-thinning fluids ($n < 1$) is given as $\alpha = 0.37$ for $n = 0.5$, and for the range of shear-thickening fluids ($n > 1$) $\alpha = 0.64$ for $n = 1.5$. The stress behaves asymptotically as $r^{-0.31}$ and $r^{-0.54}$ in the former and latter cases, respectively.

Interestingly, enough similar behavior is observed in suspensions of rigid rods aligned with the flow as pointed out by Keiller and Hinch [13]. This is important as suspension flow shows a number of similarities with inelastic and viscoelastic flows, another prominent example of which is the secondary flows of concentrated suspensions (effective concentration $\phi > 20\,\%$) in tubes of non-circular cross-section (see Siginer [14], Sect. 7.1). Keiller and Hinch [13] found that $\alpha = 0.58$ for $\phi = 5$ and $\alpha = 0.62$ for $\phi = 20$ with the stress behaving like $r^{-0.42}$ and $r^{-0.38}$, respectively. The situation is far more intractable with viscoelastic fluids and poses a much greater challenge than viscous Newtonian and inelastic generalized Newtonian fluids. However, some analytical results have been obtained in the literature for relatively simple constitutive structures valid only for small deviations from Newtonian behavior. For example, the analysis of Huilgol and Tanner [15] shows that the stress singularity in the flow of a second-order viscoelastic fluid (see Sect. 2.4.5, (2.29)) behaves very nearly like $r^{-n/(n+1)}$ with the exponent equal to $n = 0.711$. They show that shear stresses at the wall and normal stresses behave like $r^{-0.289}$ and $r^{-0.578}$, respectively.

In computational Rheology, numerical algorithms fail to converge with mesh refinement in flow domains with singularities when the working fluid is viscoelastic. With increasing values of the Weissenberg number, numerical solutions can only be obtained on increasingly coarser meshes. The current view holds that globally accurate solutions to viscoelastic flow problems in non-smooth domains can only be arrived at through a combination of numerical and analytical methods. For instance, the stress singularity in tube flows near a re-entrant corner is the focus of ongoing investigations to define the flow field around it. The predictions for the structure of the field would change with the CE used to characterize the response of the fluid. However, qualitative features of the singularity should be common to all CEs. The numerical results of Lipscomb et al. [16] are particularly significant. They seem to indicate that with increasing mesh refinement in the vicinity of the singularity, the limiting value of the Weissenberg number beyond which solutions can no longer converge decreases. This finding has been upheld by other researchers. Efforts have been made by Renardy [17, 18], Davies and Devlin [19], and others with no definitive conclusion because all analyses seem to get snagged in some sort of difficulty at some point in the development of the theoretical arguments. The issues of paramount importance here is to put on a firm footing the mathematical understanding of the flow behavior in stress boundary layers and around singularities, which may if accomplished greatly reduce the required computer power.

Clearly more computer power is not the answer to overcome these difficulties. Mesh refinement coupled with computer power is not the answer either not only due to the difficulties described above but also due to short wave instabilities of the Hadamard type stemming from the structure of the particular CE in use if the CE is subject to Hadamard type of instabilities. If the system of PDFEs and the prescribed initial and boundary conditions has a unique solution which depends continuously on the data, the problem is termed to be well posed. The problem is termed to be evolutionary if the Cauchy problem is well posed. If that is the case, then the system of PDFEs is stable to short-wave disturbances. If that is not the case, the system is no longer stable to short-wave disturbances, the growth rate of short waves increases without bound as the wavelength tends to zero and the evolution is lost. This is called a Hadamard type of instability. If evolution is lost, mesh refinement in numerical solutions will no longer improve convergence.

References

1. Tanner RI. Engineering Rheology (revised edition). Oxford: Clarendon; 1988.
2. Laun HM. Description of the non-linear shear behavior of a low density polyethylene melt by means of an experimentally determined strain dependent memory function. Rheol Acta. 1978;17(1):1–15.
3. Ferry JD. Viscoelastic properties of polymers. 3rd ed. New York: Wiley; 1980.
4. Joseph DD, Renardy M, Saut JC. Hyperbolicity and change of type in the flow of viscoelastic fluids. Arch Ration Mech Anal. 1985;87:213–51.
5. Joseph DD, Saut JC. Change of type and loss of evolution in the flow of viscoelastic fluids. J Non-Newton Fluid Mech. 1986;20:117–41.
6. Joseph DD. Fluid dynamics of viscoelastic liquids. New York: Springer; 1990.
7. Owens RG, Phillips PN. Computational rheology. London: Imperial College Press; 2002.
8. Fix GJ, Gulati S, Wakoff GI. On the use of singular functions with finite element approximations. J Comput Phys. 1973;13:209–23.
9. Dean WR, Montagnon PE. On the steady motion of a viscous liquid in a corner. Proc Camb Phil Soc. 1949;45(3):389–94.
10. Michael DH. The separation of a viscous liquid at a straight edge. Mathematika. 1958;5 (1):82–4.
11. Pearson JRA, Richardson SM, editors. Computational analysis of polymer processing. London: Applied Science Publications; 1983.
12. Henricksen P, Hassager O. Corner flow of power law fluids. J Rheol. 1989;33(6):865–79.
13. Kieller RA, Hinch EJ. Corner flow of a suspension of rigid rods. J Non-Newton Fluid Mech. 1991;40(3):323–35.
14. Siginer DA. Dynamics of tube flow of non-colloidal suspensions. New York: Springer; 2014.
15. Huilgol RR, Tanner RI. The separation of a second-order fluid at a straight edge. J Non-Newton Fluid Mech. 1977;2(1):89–96.
16. Lipscomb GG, Keunings R, Denn MM. Implications of boundary singularities in complex geometries. J Non-Newton Fluid Mech. 1987;24(1):85–96.
17. Renardy M. The stresses of an upper convected Maxwell fluid in a Newtonian velocity field near a re-entrant corner. J Non-Newton Fluid Mech. 1993;50(2–3):127–34.
18. Renardy M. A matched solution for corner flow of the upper convected Maxwell fluid. J Non-Newton Fluid Mech. 1995;58(1):83–9.
19. Davies AR, Devlin J. On corner flows of Oldroyd-B fluids. J Non-Newton Fluid Mech. 1993;50:173–91.

Chapter 2
Constitutive Formulations

Abstract Controversy about the frame indifference principle, the concept of non-local continuum field theories, local constitutive formulations, differential constitutive equations of linear viscoelasticity, Oldroyd, K-BKZ, FENE (Finitely Extensible Non-linear Elastic) class of constitutive equations, *Smoluchowski* and *Fokker–Planck* diffusion equations, constant stretch history flows, fading memory and nested integral representations of the stress, order fluids of the integral and differential type, constitutive formulations consistent with thermodynamics, maximization of the rate of dissipation in formulating thermodynamics compatible constitutive structures, Burgers equation which is finding a gradually widening niche in applications, minimum free energy and maximum recoverable work in the case of linearized viscoelastic constitutive structures, implicit constitutive theories, which define the stress field when the viscosity depends for instance on the constitutively undetermined pressure field, and which have found new focus in applications such as elastohydrodynamic lubrication are discussed and progress made is summarized. Canonical forms of Maxwell-like constitutive differential equations and single integral constitutive equations are presented and commented on together with the Hadamard and dissipative type of instabilities they may be subject to.

Keywords Non-local stress • Local stress • Linear viscoelasticity • Non-linear viscoelasticity • *Smoluchowski* diffusion equation • *Fokker–Planck* diffusion equation • Constant stretch history • Fading memory • Nested integral stress • Order fluids • Consistency with thermodynamics • Rate of dissipation • Burgers equation • Implicit constitutive structures • Canonical forms • Maxwell-like constitutive differential equations • Single integral constitutive equations • Hadamard instability • Dissipative instability

D.A. Siginer, *Stability of Non-Linear Constitutive Formulations for Viscoelastic Fluids*, SpringerBriefs in Applied Sciences and Technology 14,
DOI 10.1007/978-3-319-02417-2_2, © The Author(s) 2014

A pervasive and central problem to Non-Newtonian Fluid Mechanics and by association also central to the analysis of secondary flows is the formulation of constitutive equations. The underpinning reason for many diverse non-linear phenomena in the flow of non-linear fluids is the capability of viscoelastic fluids to accumulate recoverable strains in the course of the forcing. Viscoelastic flows are far more complex due to the effects of fading memory, which do not exist in the case of viscous fluids. As a direct consequence, the evolution of stress under steady deformation is non-linear and linear, respectively, for most viscoelastic and viscous fluids. This aspect makes even problems posed for non-linear fluids in geometrically simple settings difficult to solve analytically and recourse has to be made to numerical approach. Of course, processing flows in industry take place in complicated geometries and under high deformation rates, which makes the problem almost intractable analytically. But these difficulties pale by comparison with the unresolved issue in Non-Newtonian Fluid Mechanics of how to formulate a universal constitutive equation.

Constitutive equations in the broader meaning of the term are required to close the balance and conservation equations of mathematical physics. The balance laws in continuum physics are expressed in integral form as:

$$\frac{d}{dt}\int_{\Sigma}\mathbf{F}^0(\mathbf{x},t)\ d\Sigma = -\int_{\partial\Sigma}\mathbf{G}^i n_i d\partial\Sigma + \int_{\Sigma}\mathbf{f}\,d\Sigma,\quad \mathbf{x}\in\Sigma,\quad t\in R^+,\quad \mathbf{G}^i\in R^N\, i=1,2,3$$

which represents the rate of change of the given vector of densities $\mathbf{F}^0\in R^N$ in terms of the fluxes of quantities $\mathbf{G}^i$ through the surface $\partial\Sigma$ of the closed volume with unit normal n_i and velocity u_i. Regularity assumptions allow the above system to be expressed in differential form as:

$$\frac{\partial\mathbf{F}^0}{\partial t}+\frac{\partial\mathbf{F}^i}{\partial x^i}=\mathbf{f},\quad \mathbf{F}^i=\mathbf{F}^0 u^i+\mathbf{G}^i$$

In continuum mechanics, the balance laws, the mass conservation, the linear momentum balance, and the energy balance are explicitly written as:

$$\frac{\partial\rho}{\partial t}+\frac{\partial\rho u_i}{\partial x^i}=0,\qquad \frac{\partial\rho\, u_i}{\partial x^i}+\frac{\partial}{\partial x^i}\left(\rho u_i u_j - T_{ij}\right)=\rho b_j$$

$$\frac{\partial E}{\partial t}+\frac{\partial}{\partial x^i}\left(E u_i + q_i - T_{ij}u_j\right)=\rho\, b_i u_i,\quad E=\frac{\rho u^2}{2}+\rho\varepsilon$$

Here, E, ρ, $\mathbf{T}$, $\mathbf{q}$, and ε are the energy per unit volume, the mass density, the total stress tensor, the heat flux vector, and the internal energy per unit volume, respectively. The stress tensor $\mathbf{T}$ can further be decomposed into $\mathbf{T} = -p\,\mathbf{1} + \mathbf{S}$, where $\mathbf{S}$ is the extra-stress tensor and p and $\mathbf{1}$ stand for the pressure and the unit tensor, respectively. It should be noted that when $\mathbf{b}$ and r, which represent the body force field and the heat supply, vanish this set of field equations (balance laws) is called conservation laws. Clearly to have a closed system, another set of equations called *constitutive equations* relating, for example, the velocity field to the stresses is required.

Ruggeri [1] classifies *constitutive equations* as *local* and *non-local* in space and/or in time. His definition of non-locality is conceptually entirely different than non-local constitutive formulations of Eringen [2] (see Sect. 2.1) and should not be confused with Eringen's formulation of balance laws. For example, in the case of frictionless Euler fluids, the caloric and thermal equations of state $\varepsilon = \varepsilon\,(\rho, \theta)$, $p = p\,(\rho, \theta)$ expressing internal energy ε and pressure p in terms of the density ρ and temperature θ are *local* type of *constitutive equations*. When ploughed back into the balance equations, these equations of state yield a system of *hyperbolic* differential equations. On the other hand, the Navier–Stokes type of formulation for the extra-stress is given as:

$$\mathbf{S} = 2\nu\mathbf{D}^D + \mathbf{1}\lambda\,\mathrm{div}\,\mathbf{v},$$

where $\mathbf{D}^D$ stands for the traceless deviatoric part of the rate of deformation tensor $\mathbf{D}$ and ν and λ are phenomenological viscosity coefficients, and the Fourier's law of heat conduction $\mathbf{q}$ are classified as *non-local in space*. Constitutive equations with memory in which the stress depends not only on the deformation but also on the history of the deformation are classified as *non-local in time*. When Navier–Stokes formulation and the Fourier's law are substituted into the balance equations, one gets a system of differential equations of *parabolic* type in contrast with the Euler equations. If the equations for a mixture are considered Fick's diffusion laws [3] must be added to Fourier's law of heat conduction and the Navier–Stokes formulation of the stress–strain relationship to close the system thereby increasing the number of phenomenological constants embedded in the system to be solved. Again when the set of equations for the mixture is substituted into the balance equations a system of *parabolic* differential equations with second-order spatial derivatives and first-order time derivatives is obtained. Similarly, the well-known Darcy's equation [4] for flow in porous media is an approximation to the balance of linear momentum for the Newtonian fluid flowing through a rigid, porous solid body at the lowest order of the Maxwellian iteration process in a mathematical sense. In a physical sense, it is a very good approximation in capturing the behavior of the fluid in special circumstances, but fails spectacularly in other circumstances. The same can be said of the diffusion equations of Fick's which do work well under certain conditions. Fick's and Darcy's equations are only approximations and should not be conferred the status of physical laws just because they lead to remarkably good predictions in special circumstances, see Rajagopal [5].

Any constitutive equation must obey without fail two cornerstone principles, the material frame indifference principle (objectivity principle) and the entropy principle. The former states that all constitutive equations are independent of the observer or stated differently balance laws are invariant with respect to Galilean transformations, and the latter requires any solution of the full system of differential equations to satisfy the entropy balance law with a non-negative entropy production, see Coleman and Noll [6] and Müller [7], as given by:

$$\frac{\partial(\rho S)}{\partial t} + \frac{\partial}{\partial x^i}\left(\rho S u^i + \Phi^i\right) \geq 0$$

where Φ^i and S denote the entropy production (flux) and entropy density, respectively.

Ruggeri points out that Müller [8] proved in a landmark paper that the Navier–Stokes and Fourier equations violate the frame indifference principle. Truesdell [9] refuted the conclusions of Müller in vigorous terms. However, in spite of Truesdell rebuttal, the matter is far from settled. According to Ruggeri [1], the controversy since the publication of Müller's work in 1972 as to whether or not the frame indifference principle is a universal principle has spawned several attempts to modify the Fourier and Navier–Stokes equations to recover objectivity. Ruggeri [1] does not specifically address the arguments of Truesdell [9] in his refutation; however, he and others are clearly believers in the truth of the arguments of Müller [8].

Bressan [10] and Ruggeri [11] and more recently Ruggeri [1] working independently suggested a way out of the dilemma: Fourier and Navier–Stokes equations are not truly constitutive equations, but *approximations* representing the full set of *hyperbolic* equations in the *limit of Maxwellian iteration* resulting in a set of *parabolic* differential equations. The full set of equations describing the behavior of any physical system must be *hyperbolic* in agreement with the relativity principle, which holds that all disturbances propagate with finite speed. To quote Ruggeri [1] "*Fourier and Navier–Stokes equations are the first approximations of the Maxwellian iteration of the extended thermodynamics* (ET) *balance law system, and therefore they are not true constitutive equations and do not need to satisfy the frame indifference principle*." The question naturally arises as to whether or not the entropy principle, which is satisfied by the full hyperbolic problem, holds in the parabolic limit. The issue has been settled by Ruggeri [1] who proved that the entropy principle is preserved in the parabolic limit.

2.1 Non-local Constitutive Formulations

It is relevant to note at this point that all constitutive equations in use for non-linear viscoelastic fluids formulated, either from a continuum or molecular perspective, are based on the principle of local action. That is the material response at a point depends only on the conditions within an arbitrarily small region about that point. However, there exists a large class of problems in continuum mechanics that cannot be simulated and satisfactory predictions of flow behavior made on the basis of the principle of local action. To bring resolution to these problems development of non-local continuum field theories of material bodies, where non-local intermolecular attractions are important, is required. An example of striking anomaly that defies classical treatment is several decades of viscosity rise in fluids flowing in microscopic channels. In the field of electromagnetic theory, superconductivity cannot be treated by the classical field theories and in solid as well as fluid mechanics several branches of high-frequency waves are not predicted and short wavelength regions deviate grossly from experimental observations. Physicists often have recourse to atomic lattice dynamics to shed some light on these problems. From this perspective, a unified development of the field equations of non-local continuum field theories is a hitherto unexplored necessity.

The domain of the applicability of the classical field theories is intimately connected to the length and time scales. Denoting by L the external characteristic length such as wavelength and by l the internal characteristic length such as granular distance for example, classical field theories predict sufficiently accurate results in the region $L/l >> 1$. Local theories fail when $L/l \sim 1$ and recourse has to be made to either atomic or non-local theories to account for the long-range interatomic attractions. With time a factor in dynamical problems, there is a similar scale T/τ where T is the external characteristic time such as the time scale of the applied pressure gradient or other applied loads and τ is the internal characteristic length, for example, the time scale of transmission of signal from one molecule to the next. Classical theories fail when $T/\tau \sim 1$. Thus, the physical phenomenon in space–time requires non-locality and memory effects scaled by L/l and T/τ. Eringen [2], a pioneer in this field, points out that sometimes non-locality is inherent to the process like in solid state physics where the non-local attractions of atoms are prevalent, and the material is considered to consist of discrete atoms connected by distant forces from other neighboring atoms. Hence clearly, non-local field theories and lattice dynamics are related. Further, non-local formulations have the potential to provide bases to explain hitherto unexplained phenomena by classical field theories in rheology, biology, neural systems, and other fields.

Eringen [2] generalized local principles to develop natural extensions of the fundamental laws of physics to non-locality by reformulating the energy balance law in global form using the axiom of Galilean invariance and by requiring that the behavior of a material point in the body is influenced by the state of all points of the body at all past times. Classical notions posit that material points of a body form a continuum and that their behavior is governed by the relationships between physically independent objects called variables such as mass, charge, electric field, magnetic field assigned to the material point, and a set of response functions such as stress, internal energy, and heat. These relationships called *constitutive laws* are constructed based on the thermodynamic conditions of admissibility and additional postulates. The non-locality formulations of the basic laws encompass non-locality in both space and time that is memory dependence is woven into the theory. In non-local theories, the energy balance law is valid for the whole body and the state of the body at a material point is described by material functionals. This means that complete knowledge of the independent variables at all points of the body is required to describe the state of the body at each point. In practice, the Cauchy stress $\mathbf{S}$ at a material point $\mathbf{X}$ would be given in terms of a kernel K defined as an influence function with finite support, that is, it is identically zero for all points lying beyond a certain radius r, and the local stress $\boldsymbol{\sigma}$ at $\mathbf{X}'$:

$$\mathbf{S}(\mathbf{X}) = \int_B K\left(\left\|\mathbf{X} - \mathbf{X}'\right\|\right) \boldsymbol{\sigma}\left(\mathbf{X}'\right) dV\left(\mathbf{X}'\right)$$

where dV represents the infinitesimal volume around $\mathbf{X}'$.

To quote Eringen [2], "*Non-local field theories may provide mathematical methods for the exploration of many failures of classical field theories and for the discoveries of new physical phenomena and/or explanations of the old ones.*" For instance, the experimentally observed drastic change in fluid viscosity near rigid surfaces as compared to bulk viscosity in polymeric thin films can be explained only through the application of non-local constitutive formulations, see Eringen [2] and Eringen and Okada [12]. Specifically within a channel depth less than 30 nm, the measured viscosity of PS-cyclohexane (a solution of polystyrene in a non-polar solvent cyclohexane) is found to be several decades higher than the bulk viscosity, see Israelashvili [13, 14]. The viscosity measured near rigid surfaces is drastically higher than the bulk viscosity measured away from the walls. Based on observations it is conjectured that, near the surface, a thin layer of the order of 5–10 nm of polymeric fluid becomes rigid, reducing the channel depth. Eringen and Okada [12] argue that long-range intermolecular forces and molecular packing effects must be taken into account to successfully explain this phenomenon and proceed to show that this type of drainage phenomena can be fully explained by use of the non-local theory and by assuming that in very thin films microstructural effects play a dominant role meaning that the viscosity of the fluid depends on the molecular shapes and orientations. It is generally accepted that the molecules of orientable fluids, such as polymers and suspensions, undergo rather sharp orientation and shape changes in the vicinity of rigid boundaries. Eringen and Okada [12] conclude that the non-local theory they develop successfully predicts viscosity change with channel depth as well as the evolution of the thickness of very thin films on spinning disks. They point out that agreement of viscosity predictions with experiments is excellent to depths from 0 to 300 nm and advocate the use of non-local lubrication theory for squeezed films in narrow gaps of the order of several angstroms and further make the point that non-local lubrication theory is the key to an understanding of many dynamic processes involving very thin liquid films.

The inception of non-local theories is not new. But in spite of its age, the literature on non-local field theories is not extensive and is based essentially on the developments initiated by Eringen and his co-workers. A brief compendium of their contributions to the archival literature includes discussion of the non-local continuum theory of flowing media with or without microstructural effects, Eringen [15, 16], applications to turbulence, Speziale and Eringen [17], application to diffusion of gases, Demiray and Eringen [18], and applications to magnetohydrodynamics, Eringen [19]. Discussions of the theory of non-local electromagnetic fluids and application of the theory to dispersive waves in dielectric fluids, as given by

McCay and Narasimhan [20] and Narasimhan and McCay [21], respectively, are some of the important developments in the field. Another interesting application pointing to the potential of non-local theories of Eringen was highlighted by Speziale [22]. After demonstrating the inability of the popular isotropic k–ε turbulence model to describe secondary flows in turbulent tube flow of linear fluids, Speziale [22] shows that it is difficult to remedy this defect in the k–ε model by taking the anisotropic part of the Reynolds stress tensor to be a non-linear function of the mean velocity gradients. However, the non-local Stokesian fluid theory of Eringen does remedy this shortcoming of the k–ε model and yields secondary flows in qualitative agreement with experiments, see Speziale [23]. No further developments were initiated in this promising approach since the early 1980s. A full review of the advances made in non-local theories is beyond the scope of this book. Interested readers are referred to the above cited literature for an in-depth more comprehensive picture of the developments achieved in the field. Interest in the potential of non-local theories is on the rise in particular in solid mechanics, see Reddy [24] and Thai [25].

2.2 Local Constitutive Formulations

It is fair to say that no deep understanding of the nature of viscoelasticity has been reached as yet, that is no fundamental universal stress–strain relationship has been discovered. That leaves the field with more than ten popular constitutive equations (CE) in competition. Which is worse yet some of these CEs may yield acceptable results for certain types of flows by comparison with data, but they fail to fall in that category in other types of flows. This leads us to the inevitable concept of classes of flows and CEs, a situation opposite to the linear (Newtonian) Fluid Mechanics case. In the latter, Navier–Stokes equations are universal and apply to all types of flows of all homogeneous linear fluids. Most popular CEs in use at this time were developed without consideration of the relationship of constitutive formulations to thermodynamic principles with the exception of multiple integral formulations with fading memory, which in turn suffer from the lack of a unique way to specify the memory functionals. In addition to ignoring thermodynamic principles, most of the CEs of both differential and single integral type, which are able to describe in a limited way some of the observed phenomena in particular at low Deborah *De* numbers (the ratio of the characteristic relaxation time of the fluid to the characteristic processing time) or equivalently at low *We* numbers (the ratio of elastic forces to viscous forces) typical of laboratory settings, often turn out to be non-evolutionary. This limited ability to describe the observed phenomena within a relatively narrow range of strain rates usually seen in standard laboratory tests (that is at low *De*) such as start-up, steady state, and relaxation tests gets much worse at the high *De* numbers prevalent in processing flows in industry at least two orders of magnitude higher than those seen in the reported tests. At these high *De* numbers, the non-linear effects of elasticity become dominant. In addition to failing to describe the flow even qualitatively in the region of high *De* numbers almost all CEs exhibit various instabilities. Even molecular-based CEs such as those based on the reptation concept relatively successful in describing the behavior of dilute solutions and in general the behavior of concentrated solutions in the region of linear and weakly non-linear deformations yield very poor description of non-linear data and show physically unrealistic and unstable behavior in particular in numerical simulations in the high *De* number range. The reason for these instabilities is still elusive. Are they related to the real physical instabilities observed or are they inherent to the mathematical structure of the constitutive formulations? It is difficult to distinguish between the unstable behavior caused by poor modeling of the non-linear stress–strain relationships and the real experimentally observed physical instabilities these equations are supposed to predict. Some light has been shed on the issue by the recent work of Leonov and co-workers [26] who gave rigorous proof that the instabilities displayed by the two classes of viscoelastic CEs in general use, the class of differential Maxwell-like, and the class of time–strain separable single integral CEs, are not related to the physical instabilities observed in particular at high *De* numbers and are not caused by poorly formulated numerical algorithms, but rather that they are due to violations of fundamental principles in the formulation of the CEs themselves. Specifically almost all the popular CEs of differential and single integral type turn out to be either *Hadamard unstable* or *dissipative unstable* or both. Hadamard and dissipative instabilities are addressed in Sects. 2.8 and 2.9, respectively.

2.3 Different Approaches to Local Constitutive Formulations

Historically, the first comprehensive framework to derive CEs pioneered by Oldroyd [27] and further developed by prominent figures like Rivlin, Green, Ericksen, and Lodge postulated non-linear and quasi-linear relationships between the observable variables, the stress and the strain rate tensors and yielded CEs able to describe, albeit qualitatively, non-linear viscoelastic phenomena. The systematic framework developed by Oldroyd for the rheological behavior of rate type viscoelastic fluids introduced frame indifference restrictions, convective derivatives of physical quantities to obtain properly frame-invariant constitutive relations and the idea that the current state of stress in a body may depend on the deformation history of the body. He generalized the linear viscoelastic CE by writing it in tensorial form and imposing admissibility conditions. However, the method lacked a thermodynamic basis and hence could not describe important phenomena such as dynamic birefringence, non-isothermal flow, and diffusion, and the resulting equations were often non-evolutionary that is solutions grew exponentially with time describing physically non-plausible evolutions. If the theory is not consistent with the second law of thermodynamics, dissipative phenomena cannot be modeled.

But it should be noted that the principles in setting up CEs for viscoelastic fluids established by Oldroyd [27, 28] are nothing less than a quantum leap in this subject. Oldroyd type CEs remain the simplest suitable for modeling the dynamics of dilute polymeric solutions under general flow conditions. It is remarkable that a molecular basis consisting of a suspension of dumbbells connected by *infinitely* extensible springs (Hookean) in a Newtonian solvent leads to the Oldroyd-B as well. Oldroyd-B model has a finite (constant) shear viscosity and the extensional viscosity blows up at a finite extensional rate due to infinitely extensible interconnecting Hookean springs connecting the dumbbells. In addition, it predicts a zero second normal stress difference and consequently cannot be used to investigate secondary flows (see Siginer [29], Sect. 3.4). The shortcomings of the Oldroyd models led to the introduction of more sophisticated molecular based models, notably the variants FENE-P and FENE-CR of the FENE (*Finitely* Extensible Non-linear Elastic) class of models (see Sect. 2.3.2.2). For the sake of completeness, the structure of the Oldroyd-B model is reviewed below. The model relates the extra-stress tensor **S** to the rate of deformation tensor **D** through the constant shear viscosity η_0 and the characteristic relaxation and retardation times of the fluid λ_1 and λ_2, respectively.

$$\mathbf{S} + \lambda_1 \overset{\nabla}{\mathbf{S}} = \eta_0 \left(\mathbf{D} + \lambda_2 \overset{\nabla}{\mathbf{D}} \right)$$

The upper-convected derivative $\left(\overset{\nabla}{\bullet} \right)$ equal to the material derivative of $(\bullet)$ as it would appear to an observer in a frame of reference attached to the particle is defined as:

$$\overset{\nabla}{\mathbf{S}} = \frac{D\mathbf{S}}{Dt} - \mathbf{S}\nabla\mathbf{u} - (\nabla\mathbf{u})^{\mathrm{T}}\mathbf{S}$$

with $D(\bullet)/Dt$ standing in for the material derivative and the exponent (T) indicating transpose. If the Newtonian (solvent) and elastic (polymeric) contributions to the extra-stress tensor are separated as:

$$\mathbf{S} = \eta_s \mathbf{D} + \boldsymbol{\tau}$$

The elastic stress $\boldsymbol{\tau}$ will satisfy the constitutive relation:

$$\boldsymbol{\tau} + \lambda_1 \overset{\nabla}{\boldsymbol{\tau}} = \eta_p \mathbf{D}$$

The polymeric and solvent viscosities η_p and η_s are related through $\eta_0 = \eta_s + \eta_p$ and are as given below:

$$\eta_p = \eta_0 \left(1 - \frac{\lambda_2}{\lambda_1} \right), \quad \eta_s = \eta_0 \frac{\lambda_2}{\lambda_1}$$

The linear momentum balance written in terms of the elastic stress with f, $\mathbf{u}$, p and ρ representing the body force, the velocity, and the pressure fields and the density reads as:

$$\rho \frac{D\mathbf{u}}{Dt} = f - \nabla p + \eta_s \Delta \mathbf{u} + \nabla \bullet \boldsymbol{\tau}$$

It is noted that the concept of rate type materials predates Oldroyd and goes back to Maxwell [30], Boltzmann [31], and Jeffreys [32] who developed linear one-dimensional models to describe the viscoelastic response of materials in different contexts. For example, Maxwell was interested in the elastic and viscous behavior exhibited by gases. He realized that air can store energy and consequently can display viscoelastic behavior and devised the equation which bears his name. Jeffreys' interest in the viscoelastic response of the Earth's mantle led him to formulate the CE which bears his name. Modeling of rate type materials was taken up later by Leonov [33], Mattos [34], and Rajagopal and Srinivasa [35] who developed, in that order, thermodynamically consistent theories of non-linear CEs. Their ideas will be summarized in Sect. 2.5.

2.3.1 Linear Viscoelasticity

Following the ground breaking work of Maxwell [30], Boltzmann [31], and Jeffreys [32], models that can be categorized as of the rate type have been developed to describe the viscoelastic response of materials based on mechanical analogs. Early attempts to develop viscoelastic models were limited to one-dimensional CEs, which were based on the mechanistic models of springs and dashpots representing the coexistence of the two at a material point. This in turn supposes an additive elastic and viscous response at each material point. Banding together such springs and dashpots in series and parallel leads to a great variety of one-dimensional models. All such models lead to rate type models (or equivalent integral models) with the possibility of a spectrum of relaxation times for the material. It is not reasonable to expect a polymeric liquid of a broad molecular weight distribution to be characterized in terms of a single relaxation time; thus, the introduction of the generalized Kelvin–Voigt and the generalized Maxwell models, which mechanistically are conceived of dashpots and springs set either in series or in parallel. For instance, the generalized Maxwell model can have a finite number of Maxwell elements (a Maxwell element consists of a spring and a dashpot set in series) each with a different relaxation time from a countable infinity of Maxwell elements. Figure 2.1a shows the generalized Maxwell model with the individual elements set in parallel. A generalized Maxwell model with the elements set in series would be no different than the regular Maxwell model. Figure 2.1b and c shows the generalized Kelvin–Voigt model (a Kelvin–Voigt element consists of a spring and a

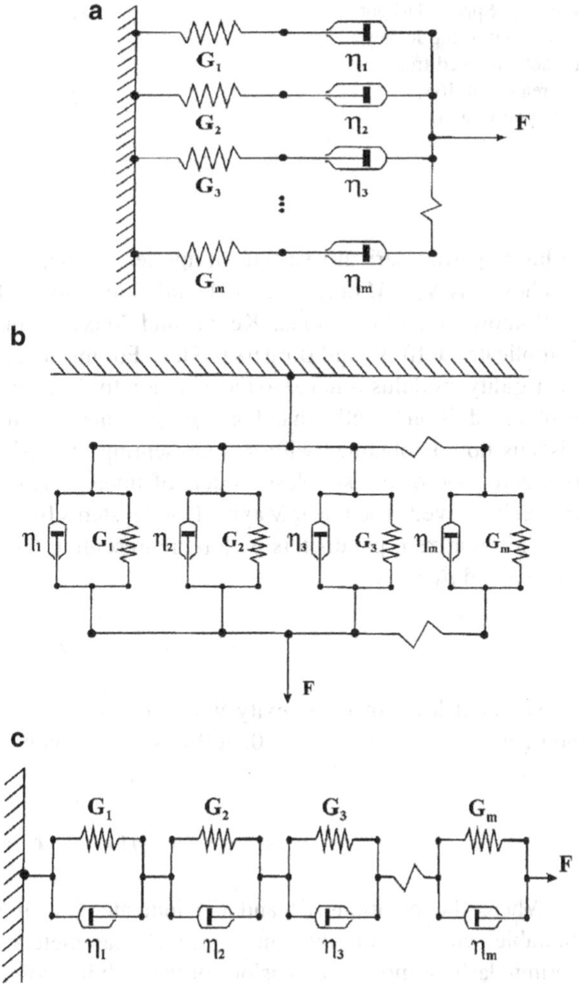

Fig. 2.1 Canonical spring-dashpot models. (**a**) Maxwell relaxation processes in parallel. (**b**) Distribution of Kelvin–Voigt retardation processes in parallel. (**c**) Distribution of Kelvin–Voigt retardation processes in series (Adapted from Carreau et al. [63]); η_i and G_i represent the viscosities and the relaxation moduli, respectively

dashpot set in parallel) with a countable set of individual elements arranged in series and in parallel.

A canonical differential equation for linear viscoelasticity can be written down as a linear relationship between the stress σ and the strain γ:

$$\left(1+\sum_{n=1}^{N}\alpha_n\frac{\partial^n}{\partial t^n}\right)\sigma=\sum_{m=0}^{M}\beta_m\frac{\partial^m}{\partial t^m}\gamma \tag{2.1}$$

The coefficients of the time derivatives α_n and β_m are constants. They are independent of the deformation measures, strain γ in this case or stresses σ, and represent material parameters such as viscosity and rigidity modulus. This equation can only entertain small changes in the variables because the time derivatives are

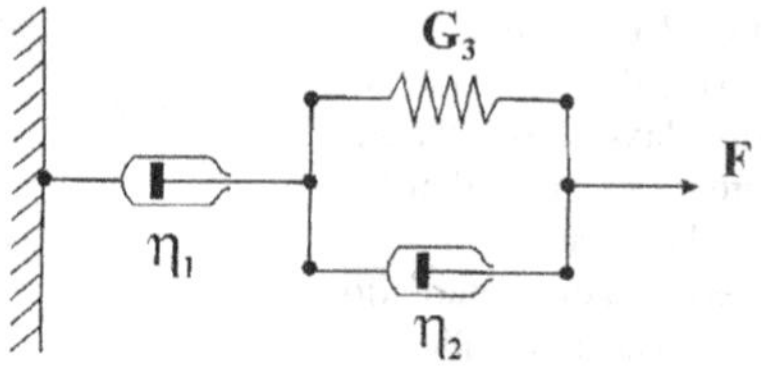

Fig. 2.2 Spring-dashpot equivalent of the Jeffreys model (Adapted from Carreau et al. [63] with permission)

ordinary partial derivatives; thus, it applies to linear viscoelasticity. N and M can be chosen as $N = M$ or $N = M - 1$. Judicious choices for the parameters yield the well-known one-dimensional Kelvin and Maxwell models as well as the more complicated Jeffreys and Burgers models. For instance, if only $\beta_0 \neq 0$ representing the rigidity modulus is not zero the equation for linearly elastic behavior (Hookean) is obtained. If on the other hand only $\beta_1 \neq 0$, the equation which defines Newtonian viscous flow is obtained with β_1 representing the viscosity. If both β_0 and β_1 are non-zero, one of the simplest models of linear viscoelasticity the Kelvin–Voight model is derived. The linear Maxwell model stems from assuming that α_1 and β_1 are the only non-zero parameters with α_1 representing the relaxation time and setting $\alpha_1 = \lambda$ and $\beta_1 = \eta$.

$$\sigma + \lambda\dot{\sigma} = \eta\dot{\gamma}$$

The next level of complexity would be assuming three of the parameters are non-zero. If $\alpha_1 \neq \beta_1 \neq \beta_2 \neq 0$, Jeffreys model with two time constants $\alpha_1 = \lambda_1$ and $\beta_2 = \lambda_2$ is obtained.

$$\sigma + \lambda_1\dot{\sigma} = \eta\left(\dot{\gamma} + \lambda_2\,\ddot{\gamma}\right)$$

Where the notations $(\dot{\ })$ and $(\ddot{\ })$ indicate first and second time derivatives. Suitable choices of the three model parameters lead to two alternative spring–dashpot models, extensions of the Kelvin–Voigt model and of the Maxwell model, which do correspond to the same mechanical behavior described by Jeffreys' model. They consist of a Kelvin–Voigt element and a dashpot in series (Fig. 2.2), and a Maxwell element arranged in parallel with a dashpot.

It is interesting to remark as a historical footnote that Jeffreys' equation was independently derived much later by Fröhlich and Sack [36] in a different context that of the flow of a suspension of elastic spheres in a linearly viscous fluid, and later by Oldroyd [37] for the flow of a dilute emulsion of an incompressible viscous liquid in another. Oldroyd [37] further showed that the effect of the interfacial slippage can be included in the simulation of the flow of a dilute emulsion if another two parameters in (2.1) are assumed to be non-zero, see Barnes et al. [38].

The well-known one-dimensional Burgers model [39, 40] involving four simple elements (two springs with different relaxation times and two dashpots with different viscosities) is obtained at the next level of complexity if four

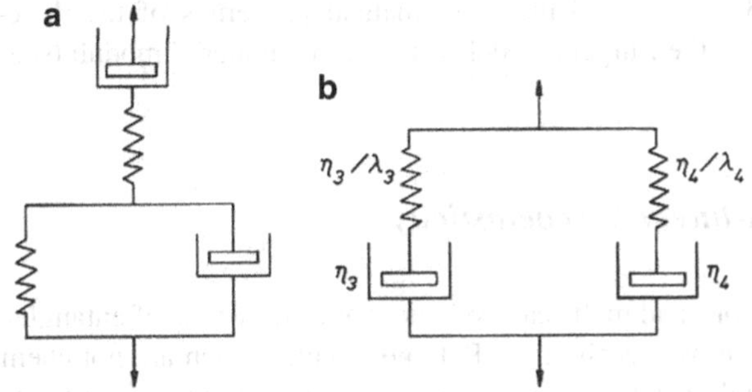

Fig. 2.3 Equivalent representations of the 4-parameter linear Burgers model; (**a**) Kelvin–Voigt and Maxwell elements in series; (b) two Maxwell elements in parallel (Reprinted from Barnes et al. [38] with permission)

parameters in (2.1) are assumed to be non-zero. One mechanical analogue of the Burgers model corresponds to a Kelvin–Voigt solid element and a Maxwell fluid element set in series as shown in Fig. 2.3a, the springs and dashpots representing energy storing and energy dissipating devices. Another mechanical analogue would correspond to two Maxwell elements set in series, see Fig. 2.3b. In terms of the parameters of the Maxwell type of mechanical representation depicted in Fig. 2.3b, the equation assumes the following form with λ_i, $i = 3, 4$ representing time constants as:

$$\sigma + (\lambda_3 + \lambda_4)\dot{\sigma} + \lambda_3\lambda_4\ddot{\sigma} = (\eta_3 + \eta_4)\dot{\gamma} + (\eta_3\ \lambda_4 + \eta_4\lambda_3)\ \ddot{\gamma} \tag{2.2}$$

There are other mechanical representations of the Burgers model, variants of the same set of two springs and two dashpots arranged differently, see Fig. 2.8a, b. This popular rate type one-dimensional viscoelastic model has been extensively used to describe the behavior of a variety of geological materials such as Earth's mantle besides viscoelastic fluids and its uses have been extended recently to asphalt and asphalt mixes by Rajagopal and his co-workers, Murali Krishnan and Rajagopal [41] and Murali Krishnan et al. [42]. The aggregate matrix in an asphalt concrete mixture has a small relaxation time, whereas the asphalt mortar matrix has relatively larger relaxation time, thus asphalt concrete mixture exhibits Burgers-like fluid behavior. The one-dimensional Burgers model is general enough to include the one-dimensional Oldroyd, Maxwell, and Navier–Stokes models as subsets. For instance, if $\lambda_3\,\lambda_4 = 0$, the Burgers model given in (2.2) reduces to the one-dimensional Oldroyd-B model. Although Oldroyd may have been aware of this, he did not consider a proper generalization of the Burgers model to three dimensions. That has been done recently by Murali Krishnan and Rajagopal [41] who used the thermodynamic framework, to systematically generate CEs for rate type viscoelastic fluids, developed by Rajagopal and Srinivasa [35] to formulate a thermodynamically compatible three-dimensional generalization of the one-dimensional Burgers model. Quintanilla and

Rajagopal [43, 44] studied the mathematical properties of the three-dimensional generalization of the Burgers model with constant material moduli (see Sect. 2.5.2).

2.3.2 *Non-linear Viscoelasticity*

Polymer solutions and melts can be looked at as networks of entangled long chains that could move over each other. Polymer chains which are not chemically cross-linked nevertheless may interact with each other to form transient networks. These temporary cross-links (constraints) do not occur at specific sites along the polymer chain but are caused by entanglements, topological constraints arising from any one given chain's inability to pass through another. Green and Tobolsky [45] originally developed network theories for solid rubber-like materials based on these concepts. The theory of rubber elasticity was extended later to polymeric melts by Lodge [46, 47], and Yamamoto [48–50] who developed semi-phenomenological theories.

This was followed by "reptation" theories due to the work of De Gennes [51] and Doi and Edwards [52]. The molecular approach has been generally successful for the viscoelastic behavior of polymeric liquids with linear or weakly non-linear deformations. However, when attempts are made to describe strong mean field flows quite arbitrary assumptions have to be made to reach plausible predictions. It is also an open question whether the rheological equations based on molecular considerations in existence in the literature are consistent with the second law of thermodynamics.

2.3.2.1 K-BKZ Type of Constitutive Formulations

The molecular theory of Doi and Edwards [52] to describe the dynamics of polymer melts is based on the idea that the motion of a molecule perpendicular to its backbone is severely constrained by surrounding molecules that form a tube along which the linear polymer chain is free to diffuse. The theory is too complicated to lend itself to analytical and even numerical investigations due to the difficulty in determining the orientation of a given segment of the polymer chain in the deformation history resulting from the coupling of different parts of the long chain as the chain retracts along the length of the tube. To make the theory accessible to computations Doi–Edwards introduced the "independent alignment" assumption, which requires that each chain segment deforms independently of the others. With this assumption Doi–Edwards model leads to an equation similar to the well-known K-BKZ (Kaye–Bernstein–Kearsley–Zapas) model (2.3), independently proposed by Kaye [53] and Bernstein et al. [54]. We note that the "independent alignment" assumption has its own limitations as shown by Marrucci and Grizzuti [55] and Marrucci [56].

$$\mathbf{T} = -p\mathbf{1} + \mathbf{S} = -p\mathbf{1} + \int_{-\infty}^{t} \left\{ \frac{\partial \phi}{\partial \mathrm{I}_{\mathbf{C}^{-1}}} (\mathrm{I}_{\mathbf{C}^{-1}}, \mathrm{I}_{\mathbf{C}}, t - t')\ \mathbf{C}^{-1}(t') + \frac{\partial \phi}{\partial \mathrm{I}_{\mathbf{C}}} (\mathrm{I}_{\mathbf{C}^{-1}}, \mathrm{I}_{\mathbf{C}}, t - t')\ \mathbf{C}(t') \right\} dt' \tag{2.3}$$

T, **S**, **1**, ϕ, and $\mathbf{C} = \mathbf{F}^{\mathrm{T}}\mathbf{F}$ stand for the total stress tensor, extra-stress tensor, the unit tensor, the potential, and the right Cauchy–Green strain tensor with **F** and $\mathbf{F}^{\mathrm{T}}$ representing the deformation gradient tensor and its transpose together with $\mathrm{I}_{\mathbf{C}^{-1}}$ and $\mathrm{I}_{\mathbf{C}}$ the first invariants, traces of $\mathbf{C}^{-1}$ and **C**, respectively, and t and t' stand for the present and past times. p represents the constitutively indeterminate part of the stress. It is remarkable that K-BKZ theory is exact in respect of the single-step strain response, and it performs reasonably well in other standard tests; however, the predictive power is not good at all for double-step strain response. A modification was proposed by Wagner [57] to remedy shortcomings of this sort by setting as:

$$\frac{\partial \phi}{\partial \mathrm{I}_{\mathbf{C}}} = 0, \quad \frac{\partial \phi}{\partial \mathrm{I}_{\mathbf{C}^{-1}}} = \mu\,(t - t')H\left(\mathrm{I}_{\mathbf{C}^{-1}}, \mathrm{I}_{\mathbf{C}}\right)$$

$$\mathbf{T} = -p\mathbf{1} + \mathbf{S} = -p\mathbf{1} + \int_{-\infty}^{t} \mu(t - t')H\left(\mathrm{I}_{\mathbf{C}^{-1}}, \mathrm{I}_{\mathbf{C}}\right)\ \mathbf{C}^{-1}(t')\, dt'$$

A different modified form of the K-BKZ equation to predict accurately the constant tensile stress and constant elongation rate experiments was suggested by Wagner [58] as:

$$\mathbf{T} = -p\mathbf{1} + \mathbf{S} = -p\mathbf{1} + \int_{-\infty}^{t} \left\{ \mu_1(\mathrm{I}_{\mathbf{C}^{-1}}, \mathrm{I}_{\mathbf{C}}, t - t')\ \mathbf{C}^{-1}(t') + \mu_2(\mathrm{I}_{\mathbf{C}^{-1}}, \mathrm{I}_{\mathbf{C}}, t - t')\ \mathbf{C}(t') \right\} dt'$$

Other modifications of the K-BKZ model are also possible by splitting the kernel into time and strain-dependent parts. These can be found in popular texts on Rheology and Non-Newtonian Fluid Mechanics, Tanner [59], Schowalter [60], Larson [61], Bird et al. [62], and Carreau et al. [63]. The examples above have been given for two reasons: K-BKZ equation, and its modified versions, which represent a high level of structural complexity remain popular in spite of their shortcomings as they pass reasonably well some standard laboratory tests such as single-step strain, double step strain, uniaxial elongation, etc. But it is glaringly obvious that, very much like in the K-BKZ example and its modification by Wagner to remedy the shortcoming of the K-BKZ in double-step strain experiment given above, the ad hoc modifications do not go beyond a band-aid effect as the altered K-BKZ that may perform acceptably well in the double-step strain experiment is quite likely to fail in the next test of higher level of complexity.

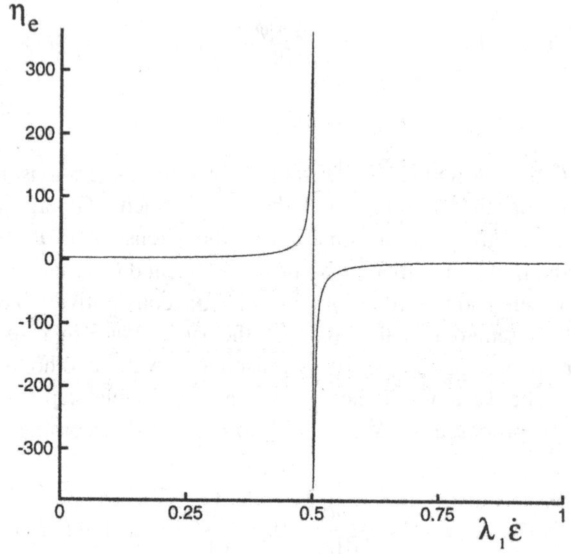

Fig. 2.4 Extensional viscosity η_e versus the characteristic relaxation time of the fluid λ_1 times the extension rate $\dot{\varepsilon}$-Oldroyd-B constitutive model (Reproduced from Owens and Phillips [81] with permission)

2.3.2.2 Constitutive Formulations of the FENE Type

Another example along these lines are the variants FENE-P and FENE-CR of the FENE (Finitely Extensible Non-linear Elastic) class of CEs, developed by Bird et al. [64], and Chilcott and Rallison [65], respectively. They were formulated to remedy the shortcomings of the Oldroyd-B equation (see Sect. 2.3). Oldroyd-B model can also be derived starting from a molecular basis consisting of a suspension of *infinitely* extensible Hookean dumbbells in a Newtonian solvent. It was natural to ask what would come out of considerations based on a more sophisticated interconnecting spring force law leading to *finite* extensibility of the interconnecting springs. One of the known deficiencies of the FENE-P model is that it does not show "stress hysteresis," see Gosh et al. [66], Larson [61]. For real polymers, two conformations with the same end-to-end distance can have different internal configurations, leading to different values of stress. In the FENE-P model, the stress depends only upon the current end-to-end distance so hysteresis is not present.

It turns out that Oldroyd-B model predicts that the extensional viscosity η_e given by:

$$\eta_e = 2\eta_0 \frac{(1 - 2\lambda_2\dot{\varepsilon})}{(1 - 2\lambda_1\dot{\varepsilon})} + \eta_0 \frac{(1 + \lambda_2\dot{\varepsilon})}{(1 + \lambda_1\dot{\varepsilon})}$$

is subject to blow-up instability at a finite extension rate $\dot{\varepsilon} = (2\lambda_1)^{-1}$. η_o is the constant shear viscosity and λ_1 represents the relaxation time, see Fig. 2.4.

The instability can be explained using the molecular basis as the dumbbells are infinitely extended at this extension rate. FENE-P model has a monotonically decreasing shear viscosity with increasing shear (shear-thinning) and bounded extensional viscosity for example. The basis of the FENE-CR model is empirical. It predicts constant shear viscosity and bounded and continuous extensional viscosities. It has been touted as a good model to simulate the flow of Boger fluids (almost purely elastic fluids with barely detectable shear-thinning if any) as there seems to be good quantitative fit with the experimental data available in the literature, Chabbra et al. [67] and Sizaire and Legat [68], and numerical simulations based on the FENE-CR model. None of these models is suitable for the study of secondary flows as they all predict a zero second normal stress difference. We note in passing that Bird and Wiest [69] is a good reference to consult for CEs for polymeric liquids.

To overcome the problem of the infinite extensibility of the Hookean spring (*linear* spring force **F** law) connecting the dumbbells, the idea of finite extensibility of the string (*non-linear* spring force **F** law) introduced by Warner [70] can be used.

$$\mathbf{F} = H\mathbf{Q}\left[1 - \left(\frac{\mathrm{tr}\mathbf{Q}\otimes\mathbf{Q}}{Q_o}\right)^2\right]^{-1}$$

With a spring law of this type the maximum length the spring can be extended is Q_o. The major issue one runs into when using a non-linear force law of the Warner type is the impossibility to arrive at a CE for the configuration *probability density function* ψ (pdf), or the configurational distribution function as it is sometimes called, directly from the diffusion (*Fokker–Planck*) equation (2.4), which can be done with the linear Hookean spring law ending up with the Oldroyd-B.

$$\frac{\partial\psi}{\partial t} = [\nabla\mathbf{u}^{\mathrm{T}}\mathbf{Q}] \bullet \frac{\partial\psi}{\partial\mathbf{Q}} - \zeta_{12}\left[kT\frac{\partial^2\psi}{\partial\mathbf{Q}^2} + \frac{\partial}{\partial\mathbf{Q}} \bullet (\psi\mathbf{F})\right] = 0 \qquad (2.4)$$

where k and T stand for a constant of proportionality and the temperature, respectively. The diffusion equation (2.4) is another form of the *Smoluchowski* equation (2.5) and is derived from it after some algebraic manipulation:

$$\frac{\partial\psi}{\partial t} = \frac{\partial}{\partial\mathbf{Q}} \bullet \left\{\nabla\mathbf{u}^{\mathrm{T}}\mathbf{Q} - \zeta_{12}\left(kT\frac{\partial}{\partial\mathbf{Q}} + \mathbf{F}\right)\right\}\psi = 0 \qquad (2.5)$$

It is worthwhile to note that the Fokker–Planck is an evolution equation, which describes the evolution of the probability density via a stochastic ordinary differential equation or a deterministic ordinary differential equation with stochastic initial values. *Smoluchowski* and *Fokker–Planck* equations are natural extensions of the *Langevin* equation. They combine macroscopic drag forces with microscopic Brownian forces, and they cannot be solved in the conventional deterministic sense.

However, if an ensemble average of dumbbells is considered and the probability ψ of finding the end-to-end vector **Q** of the dumbbell at (**x**, *t*) is sought the solution of the *Fokker–Planck* diffusion equation in a probabilistic sense is feasible. The solution of the *Fokker–Planck* equation (2.4) yields the probability ψ [**Q**(**x**, *t*),*t*] *d***Q** of finding a dumbbell with end-to-end vector **Q** in the range **Q** to **Q** + *d***Q** at (**x**, *t*). Thus, ψ (pdf) yields information on the probability of finding a dumbbell with a given configuration at a material point. The problem is *Fokker–Planck* equation is analytically intractable even for simple flows. Numerical solutions are only feasible for very low dimensional configuration space, in other words not possible at all given present-day computational resources. The alternative to attempting to solve the *Fokker–Planck* equation (2.4) is to resort to closure approximations to eliminate the ψ (pdf) and to arrive at a closed form CE for the state variables. To circumvent this difficulty, Bird et al. [64] taking their inspiration from Peterlin [71] (thus the "P" in FENE-P) made the following approximation for the spring force law to get closure and arrive at a CE for the ψ (pdf):

$$\mathbf{F} = H\mathbf{Q}\left[1 - \left\langle \frac{\mathrm{tr}(\mathbf{Q}\otimes\mathbf{Q})}{Q_0} \right\rangle^2\right]^{-1} \tag{2.6}$$

in which the spring law is preaveraged. H, **F**, **Q**, and Q_o represent some positive parameter, the force in the spring, the end-to-end vector connecting the dumbbells, and a constant, respectively. Kramers' expression for the extra-stress tensor **S**, which relates **S** to the ensemble average of the dyadic product **Q** ⊗ **F** reads as:

$$\mathbf{S} = -nkT\,\mathbf{1} + \eta_s\mathbf{D} + n\langle\mathbf{Q}\otimes\mathbf{F}\rangle$$

where *n*, *k*, and *T* represent the number density of the dumbbells, a constant of proportionality, and the temperature, respectively. The ensemble average $\langle \bullet \rangle$ of any function *f* of the dyadic product **Q** ⊗ **F** is defined in terms of the probability density function ψ (**Q**, *t*) introduced by Chandrasekhar [72]. For instance the Brownian forces generated by the solvent molecules impacting on the beads of the dumbbells are computed through:

$$\langle f(\mathbf{Q}\otimes\mathbf{F})\rangle = \iiint f(\mathbf{Q}\otimes\mathbf{F})\,\psi(\mathbf{Q},t)\,d\mathbf{Q}$$

The Giesekus form for the extra-stress tensor, sum of the Newtonian (η_s **D**) and the elastic stress tensors $\left[-\frac{n}{2\zeta_{12}}\overset{\nabla}{\langle\mathbf{Q}\otimes\mathbf{Q}\rangle}\right]$ reads as,

$$\mathbf{S} = \eta_s\mathbf{D} - \frac{n}{2\zeta_{12}}\overset{\nabla}{\langle\mathbf{Q}\otimes\mathbf{Q}\rangle}$$

$$\zeta_{12} = \zeta_1^{-1} + \zeta_2^{-1}$$

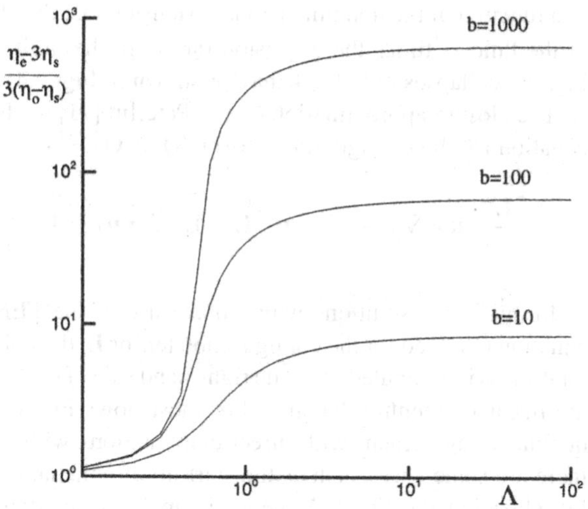

Fig. 2.5 Non-dimensional polymeric contribution to the extensional viscosity $\lambda_1\dot{\varepsilon}$- FENE-P constitutive model (Reprinted from Owens and Phillips [81] with permission)

ζ_1 and ζ_2 are the constants of proportionality (friction coefficients) arising from Stokes' law. Due to the Stokes' law the drag force on a given bead is directly proportional to the difference between the surrounding medium and the bead velocity itself. If the bead is spherical with radius a and η_s is the solvent viscosity the drag force on the ith bead is given by $\zeta_i = 6\pi\eta_s\, a_i$. Taking the upper-convected derivative of the Kramer's expression for the extra-stress tensor to eliminate $\langle \mathbf{Q} \otimes \mathbf{Q} \rangle$ between the Kramers and Giesekus forms of the elastic stress tensor, one arrives at the Oldroyd-B equation for the polymeric stress if the Hookean spring is of the simple form $\mathbf{F} = H\mathbf{Q}$. If it is of the form (2.6) the FENE-P equation is obtained.

$$Z(\mathbf{S})\,\mathbf{S} + \lambda_1 \left\{ \overset{\nabla}{\mathbf{S}} - \frac{D}{Dt}[\ln Z(\mathbf{S} + nkT\,\mathbf{1})] \right\} = nkT\lambda_1 \mathbf{D}$$

$$Z(\mathbf{S}) = 1 + \frac{3}{b}\left[1 + \frac{\mathrm{tr}\mathbf{S}}{3nkT}\right]$$

$$\lambda_1 = (2H\zeta_{12})^{-1}, \quad b = \frac{hQ_0^2}{kT}$$

where b is a dimensionless extensibility parameter. It can be shown that with increasing b shear-thinning becomes less pronounced and in the limit $b \to \infty$ a constant shear viscosity is recovered (Oldroyd-B fluid). It can also be shown that for all values of b, the extensional viscosity η_e is a continuous function of the extension rate $\dot{\varepsilon}$ in steady uniaxial extensional flow and that as $\dot{\varepsilon} \to \infty$

$$\left.\frac{\eta_e - 3\eta_s}{3\,(\eta_0 - \eta_s)}\right|_{\dot{\varepsilon}\to\infty} \quad \to \quad \frac{2}{3}(b+3) \tag{2.7}$$

where η_e, η_s and η_0 represent the extensional viscosity, the solvent viscosity, and the zero shear viscosity, respectively. Figure 2.5 shows the variation of the ratio in (2.7) as a function of the nondimensional extension rate the characteristic relaxation time of the fluid λ_1 times the extension rate $\dot{\varepsilon}$. In the limit $b \to \infty$, the trend shown in Fig. 2.5 collapses onto the behavior shown in Fig. 2.4.

The closure approximation due to Peterlin [71] leads to the following evolution equation for the configuration tensor $\langle \mathbf{Q} \otimes \mathbf{Q} \rangle = \mathbf{L}$,

$$\frac{\partial \mathbf{L}}{\partial t} + \mathbf{u} \bullet \nabla \mathbf{L} - (\nabla \bullet \mathbf{u})^{\mathrm{T}} \mathbf{L} - \mathbf{L} (\nabla \bullet \mathbf{u}) = \mathbf{1} - \frac{1}{1 - \frac{tr\mathbf{L}}{b}} \mathbf{L} = \mathbf{1} - \mathbf{S} \tag{2.8}$$

In numerical solutions which make use of the FENE-P model, this evolution equation is solved for the configuration tensor **L**, the polymeric contribution **S** to the total stress is computed from the right-hand side of (2.8) and the result is inserted into the linear momentum balance. For most flows FENE-P closure gives reasonable qualitative agreement with direct computations with the FENE theory. However, there are some cases such as the start-up of uniaxial extensional flow followed up by relaxation the FENE-P theory is unable to predict qualitatively experimental observations, which the FENE theory can. This and other cases of failure prompted a search for other, perhaps more realistic closure approximations such as the second-order closure models FENE-L, and FENE-LS (S stands for "simplified"), Lielens et al. [73, 74] who took their inspiration from Verleye and Dupret [75].

2.3.2.3 Oldroyd-B Type Constitutive Formulations

Another stream of efforts to remedy the deficiencies of the Oldroyd-B model over the years has been to add additional terms to the Oldroyd-B constitutive structure to produce shear-thinning, a second normal stress difference and a bounded extensional viscosity. All these equations can be grouped under as:

$$\mathbf{S} + \lambda \overset{\nabla}{\mathbf{S}} + \mathbf{f}(\mathbf{S}, \mathbf{D}) = \eta_p \mathbf{D}$$

where **S**, λ, η_P, and **D** represent the extra-stress tensor, the relaxation time, the polymeric contributed viscosity and the rate of deformation tensor, respectively, and $\overset{\nabla}{\mathbf{S}}$ stands for an appropriate frame indifferent convected derivative. Perhaps, the most well-known models of this family of CEs are the popular Phan-Thien–Tanner [76, 77] and Giesekus models [78, 79] (see Sect. 2.6), both of which exhibit bounded extensional viscosities, second normal stress differences, and shear-thinning. Both the Phan-Thien–Tanner and Giesekus models involve upper-convected derivatives. It should be observed that both of these models were motivated by the requirement that the extensional viscosity should increase with extension rate but remain bounded at all extension rates.

With the continuum-mechanical Oldroyd family of equations, which can be viewed as continuum-mechanical generalizations of the UCM equation, as the order of the CE is systematically increased to improve the predictive capability of the CE, the number of constants increases rapidly to pose a challenge for an accurate experimental determination of the constants. In addition, even the 8 constant Oldroyd model lacks a term that would make the extensional viscosity to increase with extension rate and yet remain bounded at all times. To produce an equation with the same predictive capability as Giesekus CE for instance, one has to go to higher orders in deriving Oldroyd type of CEs thereby increasing the number of constants further beyond 8 making the CE more unwieldy. The two versions of the Phan-Thien–Tanner [76, 77] and Giesekus [78, 79] models were introduced with the goal to make this particular constitutive feature part of the predictive capability of the CE. Giesekus equation for instance is nothing more than the UCM equation with an added quadratic term in stress (see Sect. 2.6) to insure that the extensional viscosity increase with extension rate but remain bounded at all extension rates. Although Giesekus proposed the equation that carries his name based on molecular arguments, it is remarkable that the Giesekus CE can be derived starting from a continuum perspective as well. Observation can be made that Oldroyd class of CEs are useful in modeling dilute polymeric solutions, whereas more sophisticated differential models such as the models in the FENE class of CEs, Phan-Thien–Tanner and Giesekus models to name a few of the popular differential models and integral models of the K-BKBZ class, which model a viscoelastic fluid as a relaxing rubber, are more appropriate for more concentrated solutions. Models in the K-BKBZ class can be viewed as continuum-mechanical generalizations of the integral equivalent of the UCM equation also called the Lodge equation [46, 47]. Both of these classes of models, differential and integral, account for shear-thinning and finite extensibility among other desirable features.

A major difficulty with classical constitutive models (Maxwell, Oldroyd, Giesekus, Phan-Thien–Tanner, and others) in modeling complex flows is the inability of these CEs to control independently the shear and extensional properties of the fluid. Therefore, even a qualitative description of both shear dominated and extension dominated flows is almost impossible to arrive at with a universal set of parameters. For example, with most models the viscosity can be determined using a steady simple shear flow. However, there is little flexibility left, if not none, when it comes to fitting experimental extensional data. As an example consider the UCM model. The model predicts that the extensional viscosity is given by:

$$\eta_e(\dot{\varepsilon}) = \frac{2\eta_0}{(1-2\lambda\dot{\varepsilon})} + \frac{\eta_0}{(1+\lambda\dot{\varepsilon})}$$

Note that $\eta_e \to \infty$ as $\lambda\dot{\varepsilon} \to ½$, and there are no parameters in this expression to fit experimental data, so that extensional viscosity is fixed and prescribed. In contrast, the FENE-CR model predicts,

$$\eta_e(\dot{\varepsilon}) = 3\left[\eta_e + \frac{\eta_P}{(1+\alpha)(1-2\alpha)}\right]$$

$$\alpha = \lambda\dot{\varepsilon}\left(1 - \frac{Q^2}{b}\right)$$

where Q denotes a non-dimensionalized dumbbell configuration vector. Extensional viscosity remains bounded (tends to a fixed finite value) when $\lambda\dot{\varepsilon} \to \infty$ for finite values of α, and there is a free parameter in this model to fit experimental data:

Peters et al. [80] proposed a new class of CEs to control independently the shear and extensional properties of the fluid in order to remedy the failure of some of the well-known popular CEs to model complex flows as described in the preceding paragraph. In this approach, the viscosity is considered to be a function of the relaxation time λ and a modulus G both of which are assumed to be functions of the invariants of the extra-stress tensor. For example in this approach, the generalized version of the UCM model would read as:

$$\mathbf{S} + \lambda(\mathbf{S})\stackrel{\nabla}{\mathbf{S}} = G(\mathbf{S})\lambda(\mathbf{S})\,\mathbf{D} \tag{2.9}$$

Both the relaxation time λ and the modulus G are functions of the extra-stress tensor $\mathbf{S}$ in (2.9) with the notation $\left(\stackrel{\nabla}{\bullet}\right)$ indicating upper-convected derivative. They are chosen to be functions of the invariants of the extra-stress tensor whilst maintaining the viscosity fixed. This functional dependence is loose; the only caveat being the original Maxwell model should be recovered in the limit of infinitesimal strains with the proviso that the product $G(\mathbf{S})\lambda(\mathbf{S})$ describes experimental shear viscosity data with enough accuracy. It is suggested that proposed enhanced models are better in predicting the flow structure in some of the benchmark problems such as 4:1 contraction, the cross-slot geometry, and flow past a cylinder than their unenhanced counterparts, which do fail to predict important features of the flow structure. For instance, the Phan-Thien–Tanner and Giesekus models fail to predict maximum stress levels and stress relaxation in areas with high planar elongational strain observed in the experiments, Owens and Phillips [81]. Peters et al. [80] fixed the viscosity and determined the parameters in the enhanced models from a cross-slot experiment measuring the velocities and stresses by particle tracking velocimetry and birefringence method, respectively, and used the enhanced models to predict the flow in a 4:1 contraction and flow past a cylinder. In both cases, numerical simulations with the enhanced Phan-Thien–Tanner and Giesekus models were better than the numerical simulations based on their unenhanced counterparts.

2.4 Constitutive Equation Formulations Based on Rational Continuum Mechanics

The rational continuum mechanics approach to formulating rheological equations of state, developed mainly by mathematicians starting in the 1940s by the likes of Reiner, Rivlin, and Truesdell was continued and put on solid mathematical foundation in the 1950s and following decades with the milestone contributions of Noll and Coleman, Truesdell and their colleagues, Truesdell and Noll [82]. It is no exaggeration to say that the dawn of a new era in continuum mechanics started with the papers of Reiner [83–85] and Rivlin [86–92]. A general approach to non-linear CEs was suggested by the former for the first time, and Rivlin was the first to obtain exact solutions to physical non-linear problems with response to arbitrary deformations. Both pioneers considered not only problems in non-linearly viscous fluids but in finitely strained non-linear elastic materials as well, which is the hallmark of continuum mechanics from the point of view of unified field theories. As Truesdell and Noll write in [82] "By 1949 all work on the foundations of Rheology done before 1945 had been rendered obsolete." Revolutionary ideas were introduced of which the occurrence of the second-order normal stress effects in both non-linear fluid and solid mechanics is a good example.

2.4.1 Constant Stretch History Flows

The Reiner–Rivlin CE is the most general frame-invariant constitutive equation for steady extensional flows. It is the most general CE for isotropic incompressible fluids in which the stress does not depend on the time derivatives of the velocity gradient, but depends on the instantaneous velocity gradient alone.

$$\mathbf{S} = 2\eta_1\mathbf{D} + 2\eta_2\mathbf{D}^2$$
$$\eta_i = \eta_i(\mathrm{II}_\mathbf{D}, \mathrm{III}_\mathbf{D}), \quad \mathrm{II}_\mathbf{D} = 2\mathbf{D} : \mathbf{D}, \quad \mathrm{III}_\mathbf{D} = \det\mathbf{D}$$

S and **D** are the extra-stress and rate of deformation tensors, respectively. The constitutive parameters η_i are functions of the second $\mathrm{II}_\mathbf{D}$ and third $\mathrm{III}_\mathbf{D}$ invariants of the rate of deformation tensor **D**. Steady extensional flows are part of a larger class of flows termed "motions of constant stretch history." Viscometric motions, in which all particles (material elements) have steady simple shearing histories, are the other important member of this family of motions with constant stretch history. A fluid element is undergoing a *motion with constant stretch history* if the element experiences a steady velocity gradient in a special frame that could be rotating with respect to the laboratory frame of reference. In the rotating frame, the particle sees a constant velocity gradient, but in the laboratory frame the motion may not be steady, that is the velocity gradient in the laboratory frame may be time dependent, whereas it is constant in the rotating frame. The flow is a *constant stretch history flow* if all the particles undergo motions with a constant stretch history. It is important to realize that the shear rate may vary from material element to material element in a viscometric flow; however, the history of the deformation is the same for all particles. Different flow configurations in which particle pathlines may be straight or curved or otherwise of different geometrical pattern may produce constant deformation history flows (viscometric flows). Some simple viscometric flows are flow in a cone and plate geometry and flow in the annulus between a stationary inner cylinder and rotating outer cylinder or vice versa. Some of the viscometric flows are not easy to envision; however, all viscometric flows have a common feature, shearing material surfaces that slip past each other. These surfaces are everywhere orthogonal to the velocity gradient, they may bend however they do not stretch with the distance between any two surfaces staying constant at all times even though they may not be parallel. The stress tensor for any viscometric flow has the same four components, the shear stress, and the three normal stresses. This is because in an appropriately rotated reference frame the components of the stress tensor will always be the *same four components* because in all viscometric flows all fluid elements experience a steady shearing history. Experimentally, the shear stress $S_{12} = f(D_{12} = \dot{\gamma})$ and only the differences in the normal stresses can be measured, the first normal stress difference $N_1 = S_{11} - S_{22}$ and the second normal stress difference $N_2 = S_{22} - S_{33}$. These are called the three viscometric functions.

Although in a rotated laboratory frame of reference the stress tensor is a function of S_{12}, N_1 and N_2 alone in an arbitrary frame the stress tensor in a viscometric flow is given by Criminale et al. [93] as:

$$\mathbf{S} = 2\eta\mathbf{D} - \Psi_1 \overset{\triangledown}{\mathbf{D}} + 4\Psi_2\mathbf{D}^2$$

$$\eta = \frac{S_{12}}{D_{12}}, \quad \Psi_1 = \frac{N_1}{D_{12}^2}, \quad \Psi_2 = \frac{N_2}{D_{12}^2}$$

$$\mathbf{S} = 2\eta\mathbf{D} - \Psi_1 \overset{\circ}{\mathbf{D}} + (2\Psi_1 + 4\Psi_2)\mathbf{D}^2$$

This is the CEF (Criminale–Ericksen–Filbey) CE named after their discoverers [93]. $\mathbf{S}, \mathbf{D}, \eta, \Psi_1, \Psi_2, \left(\overset{\triangledown}{\bullet}\right)$ and $\left(\overset{\circ}{\bullet}\right)$ in the above stand for the total stress tensor, the rate of deformation tensor, the viscosity, the phenomenological coefficients, the first and second normal stress coefficients Ψ_1 and Ψ_2, the upper-convected derivative, and the corotational time derivative, respectively.

Mathematically rigorous necessary and sufficient conditions to determine if the velocity field can be reduced to a motion of constant stretch history were established by Wang [94]. Clearly for steady extensional flows, the description of the flow requires only the rate of deformation tensor $\mathbf{D}$ and its square $\mathbf{D}^2$; if one goes a degree of complexity higher the description of viscometric flows will require either the upper-convected derivative $\overset{\triangledown}{\mathbf{D}}$ or the corotational derivative $\overset{\circ}{\mathbf{D}}$. If we consider another degree of complexity higher, planar motions of constant stretch history, the same tensors and the convected derivatives are sufficient to provide a description of the flow, however the phenomenological coefficients are not the same, Larson [95]. The general structure of the CEF equation stays the same with added dependence of the phenomenological coefficients on a parameter called "flow strength coefficient" S_t.

$$\mathbf{S} = 2\eta(\mathrm{II}_{\mathbf{D}}, S_t)\mathbf{D} - \Psi_1(\mathrm{II}_{\mathbf{D}}, S_t)\overset{\circ}{\mathbf{D}} + [2\Psi_1(\mathrm{II}_{\mathbf{D}}, S_t) + 4\Psi_2(\mathrm{II}_{\mathbf{D}}, S_t)]\mathbf{D}^2 \tag{2.10}$$

$$S_t = 2\frac{\left(tr\mathbf{D}^2\right)^2}{\left(\overset{\circ}{\mathbf{D}}\right):\left(\overset{\circ}{\mathbf{D}}\right)}$$

Flow classification as laid out by Tanner and Huilgol [96] requires that for strong constant stretch flows $S_t > 1$, for weak flows $S_t < 1$. In the former case, at least one eigenvalue of the velocity gradient has a positive real part, and in the former it has none. The former implies that material lines are stretched in time (exponentially), whereas the latter means that material lines oscillate in time. In the third case, when $S_t = 1$ material lines grow monotonically in time, but not exponentially. These considerations imply that almost all three-dimensional flows are strong flows. Steady elongational flow differs from steady shear flow in that some material lines may grow exponentially in time, but in steady shear flow they grow linearly in time. Thus, according to this classification, steady elongational flows are termed strong flows, whereas shear flows are weak flows. If now we consider one higher level of complexity and look at on how to represent constitutively *axisymmetric constant stretch flows*, the representation becomes a lot more complex than (2.10), and perhaps this line of inquiry describing the behavior of subclasses of motions should cautiously end at this level in favor of developing a more general approach.

2.4.2 *Simple Fluids*

The rational continuum mechanics approach was the first theory predicated on a solid thermodynamic basis and thus was able to describe the properties of all viscoelastic liquids by a set of invariant and thermodynamically consistent hereditary functionals with "fading memory," which implies that the properties of the material after it has been unloaded will be the same as they were before it was loaded if stresses or pressures applied are not large enough to cause irreversible change. The response of a simple material to any deformation history is determined by the totality of its responses to pure stretch histories. In particular, the total volumetric deformation has to be reversible if the medium has fading memory regardless whether the compressibility of the polymeric medium has been taken into account or not, Leonov [33]. There is no doubt that rational continuum mechanics approach has been very influential in formulating CEs for non-linear viscoelasticity. Given their importance the milestone developments it spawned will be reviewed next.

2.4.2.1 Fading Memory and Nested Integral Representations of the Stress

Truesdell and Noll [82] termed "simple fluids" those in which the present stress is determined by the history of the first spatial gradient of the deformation function.

However, this is too general a principle to be of predictive value and it is relaxed through the adoption of the principle of "*fading memory*" which states that "*deformations that occurred in the distant past should have less influence in determining the present stress than those that occurred in the recent past,*" which is another way of saying that the material would behave like a solid for sufficiently rapid deformations and like a Newtonian fluid for sufficiently slow deformations. The principle of fading memory can be interpreted as a requirement of smoothness for the stress response functional $\mathfrak{I}$ (see 2.12 and 2.13). The rate at which memory fades is characterized by the norm (2.11) of the linear function space of histories of the deformation $\mathbf{G}(s)$. In computing the norm the values of $\mathbf{G}(s)$ are weighted by a weight function $h(s)$ which decays with receding time into the distant past.

$$\|\mathbf{G}(s)\|_h = \left(\int_0^\infty [h(s)|\mathbf{G}(s)|]^2 ds\right)^{1/2} \tag{2.11}$$

$$|\mathbf{G}(s)| = \sqrt{tr\left[\mathbf{G}(s)^2\right]}$$

The collection of all histories with finite recollection of the above type forms a Hilbert space. It is worthwhile commenting that two histories may or may not differ very little in norm depending on whether their values are close to each other for small s (recent past) even though they may be far apart for large s (distant past). The influence of the history of the deformation in the distant past on the present deformation state depends on how fast the weight function $h(s)$ fades as time recedes into the distant past.

The constitutive behavior of simple fluids is governed by the principles of causality, material objectivity, and local action. A better light may be shed on the idea of local action in a "simple fluid" and indeed in all fluids if one considers that in the viscometric flow of a "simple fluid" viscosity is governed only by the local velocity gradient and higher velocity gradients are assumed to be inconsequential in this respect. An underlying precept of the principle of "simple fluid" is that the microstructure in all types of flows is assumed to be small enough compared with the distance over which the state of the stress changes. These assumptions allow the use of the right Cauchy–Green strain tensor $\mathbf{C} = \mathbf{F}^{\mathrm{T}}\mathbf{F}$ to describe the history of the deformation of the particle rather than using the complete description of the deformation history of all the particles in the material body to predict the future deformation of the neighborhood of a given material particle. The general isotropic, tensor valued tensor response functional $\mathfrak{I}$ of a simple fluid may be written as:

$$\mathbf{T} = -p(\rho)\mathbf{1} + \mathop{\mathfrak{I}}_{s=0}^{\infty}[\mathbf{G}(\mathbf{X}, s), \rho(t)]$$

$$\mathbf{G}(\mathbf{X}, s) = \mathbf{C}_t(t - s) - \mathbf{1}, \quad \mathbf{C}_t = \mathbf{F}_t^{\mathrm{T}}\mathbf{F}$$

where $p(\rho)$ is a scalar function of the density ρ and $\mathbf{T}$, $\mathbf{C}_t$, and $\mathbf{F}_t$ are the total stress, the relative right Cauchy–Green strain tensor, and the relative deformation gradient, respectively with $\mathbf{G}(\mathbf{X}, s)$ representing the history of the strain on the particle located at $\mathbf{X}$ at rest. For incompressible simple fluids, this expression reduces to:

$$\mathbf{T} = -p\mathbf{1} + \underset{s=0}{\overset{\infty}{\mathfrak{I}}}[\mathbf{G}(\mathbf{X}, s)] \tag{2.12}$$

or equivalently,

$$\mathbf{T} = -p\mathbf{1} + \underset{t'=-\infty}{\overset{t'=t}{\mathfrak{I}_1}}[\mathbf{C}(\mathbf{X}, t')] \tag{2.13}$$

The response functional $\mathfrak{I}$ satisfies the isotropy relation:

$$\mathbf{Q}\left(\underset{s=0}{\overset{\infty}{\mathfrak{I}}}[\mathbf{G}(\mathbf{X}, s)]\right)\mathbf{Q}^{\mathrm{T}} = \underset{s=0}{\overset{\infty}{\mathfrak{I}}}[\mathbf{Q}\mathbf{G}(\mathbf{X}, s)\mathbf{Q}^{\mathrm{T}}] \quad \forall \mathbf{Q}$$

for all orthogonal tensors $\mathbf{Q}$. As it stands not much can be deduced from (2.12) and (2.13) except certain symmetries of $\mathbf{T}$ given the symmetry of $\mathbf{C}_t$. Green and Rivlin [97, 98] and Green et al. [99] in a series of papers developed an integral expansion of the functional $\mathfrak{I}$ for finite strains, which led to the integral type of fluids of order n. They assume that a suitable function space endowed with a suitable topology can be assigned to the domain of the response functional $\mathfrak{I}$, and that $\mathfrak{I}$ is continuous with respect to that topology. Given these premises, Stone–Weierstraβ theorem can be used to show that the response functional $\mathfrak{I}$ can be uniformly approximated by integral polynomials, that is the response functional can be expanded into a uniformly approximating series of multiple integral terms with tensor polynomial integrands as:

$$\underset{s=0}{\overset{\infty}{\mathfrak{I}}}[\mathbf{G}(\mathbf{X}, s)] = \sum_{1}^{\infty} \mathbf{S}_n = \int_0^\infty \mathbf{K}_1(s)\,\mathbf{G}(s)\,ds + \int_0^\infty\!\!\int_0^\infty \mathbf{K}_2(s_1, s_2)\,\mathbf{G}(s_1)\mathbf{G}(s_2)\,ds_1 ds_2$$
$$+ \int_0^\infty\!\!\int_0^\infty\!\!\int_0^\infty \mathbf{K}_3(s_1, s_2, s_3)\,\mathbf{G}(s_1)\mathbf{G}(s_2)\mathbf{G}(s_3)\,ds_1 ds_2 ds_3 + \cdots \tag{2.14}$$

where the even order kernel tensors $\mathbf{K}_i$ ultimately define the material functions and therefore the behavior of the fluid. For isotropic liquids the integrands are isotropic tensor polynomials and the $\mathbf{K}_i$ are isotropic tensors of even order. Mathematically manageable forms of the stress response functional $\mathfrak{I}$ can be obtained if $\mathfrak{I}$ is linearized around some deformation history $\mathbf{G}_o$ assuming functional differentiability of either Fréchet or Gateaux type.

$$\mathop{\mathfrak{J}}_{s=0}^{\infty}[\mathbf{G}(\mathbf{X},s)] = \mathfrak{J}_0[\mathbf{G}_0] + \delta\mathfrak{J}[\mathbf{G}_0|\ \mathbf{G}_{00}] + \delta^2\mathfrak{J}[\mathbf{G}_0|\ \mathbf{G}_{00},\mathbf{G}_{00}]$$
$$+ \delta^3\mathfrak{J}[\mathbf{G}_0|\ \mathbf{G}_{00},\mathbf{G}_{00},\mathbf{G}_{00}] + O\left(|\mathbf{G}_{00}|^4\right)$$

where $\delta^{(i)}\mathfrak{J}$ represent functional derivatives evaluated at $\mathbf{G}_o$ and the history of the particle is expressed as the sum of the history of the base state $\mathbf{G}_o$ and the history of the deviation $\mathbf{G}_{oo}$ from the base state.

$$\mathbf{G}(\mathbf{X},s) = \mathbf{G}_0(\mathbf{X},s) + \mathbf{G}_{00}(\mathbf{X},s) \tag{2.15}$$

Any functional representation would mathematically make sense only in a suitable function space when the functional representation is continuous with respect to an appropriate continuity measure in that space. The topological structure of the chosen space determines the behavior of the fluid. Under different topologies, the fluid will behave differently in each and every topology. The assumed topology determines the structure of the space and defines the domain, the class of admissible deformation histories, which in turn defines the range of the response functional the collection of all possible stresses under the assumed topology. The continuity of the response functional under the assumed topology defines in what way stresses at the present time are dependent on the strains the material has been subjected to in the past. The principle of "fading memory" which defines how strongly the present stresses depend on the recent deformation history and how weakly the material remembers the effect of the imposed strains in the distant past was introduced earlier in this section. The theories in this section live in the rapidly decaying weighted fading memory norm (2.11) introduced by Coleman and Noll [100], and reviewed by Truesdell and Noll [82].

The structure of (2.14) may be better understood physically if we consider that the single integral term represents the effect on the stress of the strain increments added at various times in the past, and that the effect of an increment at one time has no influence on the effect of another increment at a later time. The double and triple integral terms estimate the contribution to the stress from two and three contributions to the strain added at different times, respectively.

Theories of fading memory have been formulated by Wang [101], Coleman and Mizel [102], and Saut and Joseph [103] besides Coleman and Noll [100]. A theory of fading memory is strongly dependent on the measure of continuity appropriate to the space in which the constitutive relationship is valid. Coleman and Noll use a Hilbert space with a rapidly decaying fading memory norm. The domain of the response functional in their formulation admits a large class of deformation histories some of which may not be smooth. Coleman and Noll's theory allows shocks, whereas Saut and Joseph's theory [103] does not.

2.4.2.2 Order Fluids of the Integral Type

The response functional $\Im_N$ of the integral fluid of Nth order is obtained if the series (2.14) are truncated after the Nth term. The CE for an incompressible fluid of the first order reads as:

$$\mathbf{T} = -p\mathbf{1} + \mathbf{S}_1, \quad \mathbf{S}_1 = \int_0^{\infty} \zeta(s)\,\mathbf{G}(s)\,ds \tag{2.16}$$

and the CE for an incompressible second-order fluid of the integral type is given by:

$$\mathbf{T} = -p\mathbf{1} + \mathbf{S}_1 + \mathbf{S}_2$$

$$\mathbf{S}_2 = \int_0^{\infty}\int_0^{\infty} \beta_{21}[(s_1, s_2)\; tr\mathbf{G}(s_1)\;\; \mathbf{G}(s_2) + \beta_{22}(s_1, s_2)\;\mathbf{G}(s_1)\mathbf{G}(s_2)]\;\; ds_1 ds_2 \tag{2.17}$$

Whereas the integral fluid of order three reads as:

$$\mathop{\Im_3}_{s=0}^{\infty}[\mathbf{G}(\mathbf{X}, s)] = \sum_1^3 \mathbf{S}_n, \quad \mathbf{S}_3 = \sum_1^4 \mathbf{S}_{3i} \tag{2.18}$$

$$\mathbf{S}_{31} = \int_0^{\infty}\int_0^{\infty}\int_0^{\infty} \beta_{31}(s_1, s_2, s_3)\;\mathbf{G}(s_1)\mathbf{G}(s_2)\mathbf{G}(s_3)\;\; ds_1 ds_2 ds_3 \tag{2.19}$$

$$\mathbf{S}_{32} = \int_0^{\infty}\int_0^{\infty}\int_0^{\infty} \beta_{32}(s_1, s_2, s_3)\; tr\mathbf{G}(s_1)\;\; \mathbf{G}(s_2)\mathbf{G}(s_3)\;\; ds_1 ds_2 ds_3 \tag{2.20}$$

$$\mathbf{S}_{33} = \int_0^{\infty}\int_0^{\infty}\int_0^{\infty} \beta_{33}(s_1, s_2, s_3)\; tr\mathbf{G}(s_1)\;\; tr\mathbf{G}(s_2)\mathbf{G}(s_3)\;\; ds_1 ds_2 ds_3 \tag{2.21}$$

$$\mathbf{S}_{34} = \int_0^{\infty}\int_0^{\infty}\int_0^{\infty} \beta_{34}(s_1, s_2, s_3)\; tr[\mathbf{G}(s_1)\;\; \mathbf{G}(s_2)]\,\mathbf{G}(s_3)\;\; ds_1 ds_2 ds_3 \tag{2.22}$$

Where the unknown kernel functions $\zeta(s)$, $\beta_{21}(s_1, s_2)$, $\beta_{22}(s_1, s_2)$, $\beta_{31}(s_1, s_2, s_3)$, and $\beta_{32}(s_1, s_2, s_3)$, $\beta_{33}(s_1, s_2, s_3)$, β_{34} (s_1, s_2, s_3) are material-dependent functions. The strain history $\mathbf{G}(\mathbf{X}, s)$ in (2.15) may be given a more explicit form if the strain history deviation $\mathbf{G}_{oo}$ is expanded in a power series in terms of a small physical parameter ε:

$$\mathbf{G}(\mathbf{X}, s) = \sum_0^{\infty} \varepsilon^n \mathbf{G}_n(\mathbf{X}, s)$$

Further assuming that the response functional is analytic with respect to ε to allow the existence of functional derivatives of the Fréchet type and following Joseph [104] an approximation to the response functional can be constructed as a Fréchet series pivoted around the base state $\mathbf{G}_o$:

$$\begin{aligned}\underset{s=0}{\overset{\infty}{\mathfrak{I}}}[\mathbf{G}(\mathbf{X},s)] &= \underset{s=0}{\overset{\infty}{\mathfrak{I}}}\left[\sum_{0}^{\infty}\varepsilon^n\mathbf{G}_n(\mathbf{X},s)\right] = \mathfrak{I}_0[\mathbf{G}_0] + \mathfrak{I}_{,\varepsilon}[\mathbf{G}_0{:}\varepsilon\mathbf{G}_1] \\ &+ \frac{1}{2!}\mathfrak{I}_{,\varepsilon\varepsilon}\left[\mathbf{G}_0{:}\varepsilon^2\mathbf{G}_2,\varepsilon\mathbf{G}_1,\varepsilon\mathbf{G}_1\right] \\ &+ \frac{1}{3!}\mathfrak{I}_{,\varepsilon\varepsilon\varepsilon}\left[\mathbf{G}_0{:}\varepsilon^3\mathbf{G}_3,\varepsilon^2\mathbf{G}_2,\varepsilon\mathbf{G}_1,\varepsilon\mathbf{G}_1,\varepsilon\mathbf{G}_1\right] + \cdots\end{aligned} \tag{2.23}$$

$\mathfrak{I}_{,\varepsilon}$, $\mathfrak{I}_{,\varepsilon\varepsilon}$, and $\mathfrak{I}_{,\varepsilon\varepsilon\varepsilon}$ indicate Fréchet derivatives of the 1st, 2nd and 3rd orders, respectively, with respect to ε. Next the response functional of the integral fluid of order N can be rewritten through identifying the structure of the integral fluid of order N with the N-term Fréchet expansion of the stress response functional around the rest state $\mathbf{G}_o$ as:

$$\underset{s=0}{\overset{\infty}{\mathfrak{I}_N}}[\mathbf{G}(\mathbf{X},s)] = \sum_{1}^{N}\frac{\varepsilon^n}{n!}\frac{\partial^n\mathbf{S}}{\partial\varepsilon^n}\bigg|_{\varepsilon=0} = \sum_{1}^{N}\varepsilon^n\mathbf{S}^{(n)} \tag{2.24}$$

where the partial derivative is evaluated at $\mathbf{G}_o$ at $\varepsilon = 0$. Note that due to incompressibility:

$$\begin{gathered} tr\mathbf{G}(\mathbf{X},s) = \varepsilon^2\, tr\mathbf{G}_2(\mathbf{X},s) + O(\varepsilon^3) \\ \mathbf{S}_2 \approx O(\varepsilon^3),\quad \mathbf{S}_{32} \approx O(\varepsilon^4),\quad \mathbf{S}_{33} \approx O(\varepsilon^5),\quad \mathbf{S}_{34} \approx O(\varepsilon^3)\end{gathered} \tag{2.25}$$

The Fréchet derivatives $\mathbf{S}^{(n)}$ in (2.23) evaluated at the rest state $\mathbf{G}_o$ obtained from equations (2.16) to (2.23) above and expressed in terms of the first Rivlin–Ericksen tensor alone [see (2.26) for the definition of Rivlin–Ericksen tensors] and new kernel functions derived from the kernel functions in (2.17)–(2.22) are collected in the Appendix.

The process that leads to Fréchet stress representations in (2.24) assumes that the functional derivatives of the stress response functional evaluated at the base state can be represented as integrals. Although Riesz theorem provides a theoretical basis for the representation of the first functional derivative $\delta\mathfrak{I}$ at $\mathbf{G}_o$ as a single integral with the integrand linear in the strain history deviation $\mathbf{G}_{oo}$ representation theorems to justify the representation of the second $\delta^2\mathfrak{I}$ and third functional $\delta^3\mathfrak{I}$ derivatives at $\mathbf{G}_o$ as double and triple-nested integrals bilinear and trilinear, respectively, in $\mathbf{G}_{oo}$ do not exist, and the above representations as double and triple-nested integrals are nothing more than a constitutive hypothesis to be tested through comparison of predictions with experimental facts.

Joseph [105, 106] gave the canonical forms of the functional derivatives evaluated at $\mathbf{G}_o$ when $\mathbf{G}_o$ is either a rigid body rotation or the rest state when the small parameter ε in (2.24) represents a small amplitude perturbation. Previously, small amplitude perturbations of steady viscometric flows were investigated by Pipkin and Owen [107] who derived the canonical form of the first Fréchet derivative and consistency relationships between $\delta\mathfrak{J}$ and viscometric functions and determined that 13 elements of $\delta\mathfrak{J}$ are non-zero due to material symmetry, isotropy, and incompressibility. Zahorski [108, 109] investigated flows with proportional stretch histories, which includes nearly viscometric flows as a subclass and derived the canonical form for the first functional derivative with the same number of independent constitutive functions as Pipkin and Owen [107].

Applications of order fluids of the integral type are rare due to the difficulties in determining the constitutive parameters, which are numerous even at the third order, and to the difficulties in numerical implementation. More will be said about this issue later on in this section after the introduction of the order fluids of the differential type and differential fluids of grade n. However, successful uses exist in the literature. For example, Siginer [110, 111] used the response functional of the third-order fluid expressed in terms of the Fréchet stresses to study the non-linear effects and anomalous flows stemming from nearly viscometric flows due to longitudinal and transversal boundary vibrations superposed on the Poiseuille flow and the pulsating pressure gradient driven flow of rheologically complex fluids. From an experimental point of view, the kernel (memory function) of the first integral in (2.14) or $\zeta(s)$ in (2.16) is not difficult to determine as it turns out that it is the derivative of the linear relaxation function $G(t)$. However, higher-order kernels are not so easy to find.

In the development of the order fluids of the integral type, the relative right Cauchy–Green strain tensor $\mathbf{C}_t$ has been used to describe the deformation history. Truesdell and Noll [82] remark that any other equivalent deformation measure such as $\mathbf{C}_t^{-1}$ or the relative right stretch tensor $\mathbf{U}_t = \sqrt{\mathbf{C}_t}$ could have been used as well. They further note that "the concept of a material of the integral type depends on the choice of the deformation measure", which means that a material which is of the integral type with respect to $\mathbf{C}_t$ is not of the integral type with respect to $\mathbf{U}_t$.

The kernel $\zeta(s)$ is not difficult to determine experimentally as it is the derivative of the linear time-dependent shear relaxation function (modulus) $G(t)$ of the fluid in stress relaxation. However, higher-order kernels are not easy to determine. The results of Beavers [112], who used the quasi-unsteady flow field and deformation measurements of the free surface on the simple fluid between torsionally oscillating cylinders to determine the kernels of the second-order integral fluid, are particularly noteworthy in this regard. The physical interpretation of the single integral form in (2.16), which defines linear viscoelastic behavior, is that the integral represents the effect of strain increments added at various times in the past such that the effect of any given increment at time t_1 is independent of the effect of another increment at a later or earlier time. The possibility of such an influence is recognized in the double integral form, which estimates the contribution to the stress from two contributions to the strain added at different times.

2.4.2.3 Order Fluids of the Differential Type

In an isotropic fluid, the extra-stress tensor at a given instant t depends only on the gradients of the velocity, acceleration, and higher time derivatives at the same instant t. Rivlin and Ericksen [113] introduced in 1955 the Rivlin–Ericksen tensors $\mathbf{A}_k$ as:

$$\mathbf{A}_k(\mathbf{x},t) = \frac{\partial^k}{\partial t'^k}\mathbf{C}(\mathbf{x},t,t')\bigg|_{t'=t} \tag{2.26}$$

and used invariance arguments to show that the dependence of the extra-stress $\mathbf{S}$ on these gradients must be through combinations of $\mathbf{A}_n$, and that this dependence can be expressed as a tensor valued polynomial function f of the first N Rivlin–Ericksen tensors for a differential fluid of order N. If the fluid is isotropic, they showed that f must be an isotropic function of its arguments as:

$$\mathbf{S} = f(\mathbf{A}_1, \mathbf{A}_2, \ldots, \mathbf{A}_N)$$

They further proved that f reads as in (2.27) if $\mathbf{S}$ depends only on $\mathbf{A}_i\, i = 1, 2$, and given by:

$$\mathbf{S} = f(\mathbf{A}_1, \mathbf{A}_2) = \sum_0^2 \sum_0^2 \varphi_{mn}\left(\mathbf{A}_1^m \mathbf{A}_2^n + \mathbf{A}_2^n \mathbf{A}_1^m\right) \tag{2.27}$$

The scalar functions φ_{mn} are functions of the traces of products of $\mathbf{A}_1$ and $\mathbf{A}_2$. It is remarkable that the form (2.27) was obtained under the assumption of isotropy alone and no other assumption about the nature of the tensorial function f or about the magnitude of the deformation for instance the shear rate. In viscometric flows all Rivlin–Ericksen tensors of order higher than two are zero, and as a consequence (2.27) becomes exactly equivalent to the general CE for an incompressible simple fluid in depicting the dependence of the stress on the deformation history.

The partial derivative in (2.26) with respect to t' is taken following the particle in the Lagrangian sense that is at fixed $\mathbf{x}$ and t. All Rivlin–Ericksen tensors are symmetric and any Rivlin–Ericksen tensor $\mathbf{A}_n$ of order n has dimension t^{-n} where t is time. Higher-order tensors can be deduced from the following recurrence relation in an Eulerian frame of reference:

$$\mathbf{A}_{k+1} = \frac{D}{Dt}\mathbf{A}_k + (\nabla\mathbf{u})\mathbf{A}_k + \mathbf{A}_k(\nabla\mathbf{u})^{\mathrm{T}}, \quad \mathbf{A}_1 = 2\mathbf{D} \tag{2.28}$$

One way to derive the CE for an order fluid of the differential type is to replace $\mathbf{G}(s)$ in (2.12) or equivalently $\mathbf{C}$ in (2.13) with its expansion in a Taylor series in the $\mathbf{A}_n$ for sufficiently smooth motions:

$$\mathbf{C}(t') = \mathbf{1} - \mathbf{A}_1(t-t') + \frac{1}{2!}\mathbf{A}_2(t-t')^2 + \frac{1}{3!}\mathbf{A}_3(t-t')^3 + \cdots$$

This process leads to the CE of frictionless fluids (Euler equations) at the zeroth order and viscous Newtonian fluids at the first order:

$$\mathbf{T} = -p\mathbf{1}, \quad \mathbf{T} = -p\mathbf{1} + \eta_0\mathbf{A}_1$$

At the next order, it leads to the differential fluid of second order with constant shear rate viscosity and the first and second normal stress coefficients Ψ_1 and Ψ_2 (widely used by rheologists but not by practitioners of continuum mechanics), respectively.

$$\mathbf{T} = -p\mathbf{1} + \eta_0\mathbf{A}_1 + (\Psi_1 + \Psi_2)\mathbf{A}_1^2 - \frac{1}{2}\Psi_1\mathbf{A}_2 \tag{2.29}$$

The fluid of the third-order features a viscosity that changes with the shear rate and the relaxation modulus λ, and has additional material constants φ_i, $i = 1, 2, 3$.

$$\mathbf{T} = -p\mathbf{1} + \eta_0\left(1 + \lambda^2\varphi_1 tr\mathbf{A}_1^2\right)\mathbf{A}_1 + (\Psi_1 + \Psi_2)\mathbf{A}_1^2 - \frac{1}{2}\Psi_1\mathbf{A}_2$$
$$+ \varphi_2\lambda^3(\mathbf{A}_1\mathbf{A}_2 + \mathbf{A}_2\mathbf{A}_1) + \varphi_3\lambda^3\mathbf{A}_3$$

The major drawbacks of these models are for the second-order model an inability to simulate shear-rate dependent viscosity in addition to unsuitability to describe unsteady flows due to the onset of instabilities. In fact all unsteady flows are too fast to simulate even for order fluids of higher order as instabilities set in, Joseph [114]. Further, the third-order model sets a limit on the maximum allowable shear rate for the model to depict a realistic shear stress behavior. Beyond that limit, the third-order model does not predict a monotonically increasing shear stress with shear rate, Bird et al. [62]. These remarks are valid as well for differential fluids of grade n whose structure will be summarized next.

Coleman and Noll [100] introduced in 1960 "retarded histories" of the deformation in terms of a retardation parameter α as:

$$\mathbf{G}_\alpha(\mathbf{x}, t, s) = \mathbf{G}(\mathbf{x}, t, \alpha s), \quad 0 < \alpha < 1$$

with the corresponding modified Rivlin–Ericksen tensors:

$$\mathbf{A}_k^\alpha(\mathbf{x}, t) = \alpha^k(-1)^k \frac{\partial^k}{\partial(\alpha s)^k}\mathbf{G}(\mathbf{x}, t, \alpha s)\bigg|_{\alpha s = 0}, \quad 0 < \alpha < 1$$

When the recollection $\|\mathbf{G}(s)\|_h$ of the history in the topology defined in (2.11) is small, that is taking the present configuration of the material as reference $\|\mathbf{G}(s)\|_h$ is small in the recent past even though it may have been large in the distant past, the general constitutive equation (2.12 and 2.13) with fading memory is well approximated by writing the new extra-stress tensor $\mathbf{S}_\alpha$ as a multilinear isotropic function of the modified Rivlin–Ericksen tensors $\mathbf{A}_k^\alpha(\mathbf{x},t)$. Coleman and Noll [100] were able to show that it was possible to expand the retarded extra-stress $\mathbf{S}_\alpha$ to various orders when the strain history is retarded by a factor α to give the stress to within terms of order $O(\alpha^n)$ in the retardation factor α. These expansions are known as "the retarded motion expansions." At the first four orders they read as:

$$\mathbf{S}_1^\alpha = \eta_0 \mathbf{A}_1^\alpha + O(\alpha)$$

$$\mathbf{S}_2^\alpha = \mathbf{S}_1^\alpha + \beta \mathbf{A}_1^\alpha \mathbf{A}_1^\alpha + \gamma \mathbf{A}_2^\alpha + O(\alpha^2)$$

$$\mathbf{S}_3^\alpha = \mathbf{S}_2^\alpha + \beta_1 \mathbf{A}_3^\alpha + \beta_2\left(\mathbf{A}_1^\alpha \mathbf{A}_2^\alpha + \mathbf{A}_2^\alpha \mathbf{A}_1^\alpha\right) + \beta_1\left(tr\mathbf{A}_2^\alpha\right)\mathbf{A}_1^\alpha + O(\alpha^3)$$

$$\begin{aligned}\mathbf{S}_4^\alpha = \mathbf{S}_3^\alpha &+ \gamma_1 \mathbf{A}_4^\alpha + \gamma_2\left(\mathbf{A}_1^\alpha \mathbf{A}_3^\alpha + \mathbf{A}_3^\alpha \mathbf{A}_1^\alpha\right) + \gamma_3 \mathbf{A}_2^\alpha \mathbf{A}_2^\alpha + \gamma_4\left(\mathbf{A}_1^\alpha \mathbf{A}_1^\alpha \mathbf{A}_2^\alpha + \mathbf{A}_2^\alpha \mathbf{A}_1^\alpha \mathbf{A}_1^\alpha\right) + \gamma_5\left(tr\mathbf{A}_2^\alpha\right)\mathbf{A}_2^\alpha \\ &+ \gamma_6\left(tr\mathbf{A}_2^\alpha\right)\mathbf{A}_1^\alpha \mathbf{A}_1^\alpha + \left[\gamma_7 tr\mathbf{A}_3^\alpha + \gamma_8 tr\left(\mathbf{A}_2^\alpha \mathbf{A}_1^\alpha\right)\right]\mathbf{A}_1^\alpha + O(\alpha^4)\end{aligned}$$

where η_0 represents the zero shear viscosity and $\beta, \gamma, \beta_i, \gamma_j, i = 1, 2; j = 1, 2, \ldots, 8$ are constants. These asymptotic approximations to the response functional when a given flow with history $\mathbf{G}(s)$ is retarded are called "fluids of the differential type of grade n." It should be observed that these asymptotic expansions apply only to "slow" and "slowly varying" flows. For good predictive results, both conditions should be met. For instance, consider the flow structure near a reentrant corner in tube flow. If boundary conditions are of the no-slip type, flow in the vicinity of the corner will be slow for moderate Reynolds numbers. However, flow around the corner can never be regarded as "slowly varying" in any sense and retarded asymptotic expansions of the differential fluids of grade n type should not be used.

Rivlin [115] has shown that if $\forall \mathbf{A}_k = 0$ for $k \geq 3$, the extra-stress $\mathbf{S}$ can be expressed as a function of eight polynomials α_i, $i = 1, \ldots, 8$ in $tr\mathbf{A}_1$, $tr\mathbf{A}_2$, $tr\mathbf{A}_1^2$, $tr\mathbf{A}_1^3$, $tr\mathbf{A}_2^2$,$\text{tr}\mathbf{A}_2^3$, $tr\mathbf{A}_1\mathbf{A}_2$, $tr\mathbf{A}_1^2\mathbf{A}_2$, $tr\mathbf{A}_1\mathbf{A}_2^2$, $tr\mathbf{A}_1^2\mathbf{A}_2^2$:

$$\begin{aligned}\mathbf{S} = \alpha_0 \mathbf{1} &+ \alpha_1 \mathbf{A}_1 + \alpha_2 \mathbf{A}_2 + \alpha_3 \mathbf{A}_1^2 + \alpha_4 \mathbf{A}_2^2 + \alpha_5\left(\mathbf{A}_1\mathbf{A}_2 + \mathbf{A}_2\mathbf{A}_1\right) + \alpha_6\left(\mathbf{A}_1^2\mathbf{A}_2 + \mathbf{A}_2\mathbf{A}_1^2\right) \\ &+ \alpha_7\left(\mathbf{A}_2^2\mathbf{A}_1 + \mathbf{A}_1\mathbf{A}_2^2\right) + \alpha_8\left(\mathbf{A}_1^2\mathbf{A}_1^2 + \mathbf{A}_2^2\mathbf{A}_2^2\right)\end{aligned} \tag{2.30}$$

It turns out that is exactly the case for steady simple shearing flows $\mathbf{u} = (D_{xy}\, y, 0, 0)$. Thus (2.30) is exact for steady shearing flows. Criminale et al. [93] further simplified (2.30) and showed that it can be reduced to one involving only $\mathbf{A}_1$, $\mathbf{A}_2$, and $\mathbf{A}_1^2$. If the fluid is also incompressible $tr\mathbf{A}_1 = 0$ and the polynomial coefficients become function of $D_{xy} = \dot{\gamma}$ alone. Further inspection of the coefficients leads to writing the final expression as:

$$\mathbf{T} = -p\mathbf{1} + \eta_0(\dot{\gamma})\,\mathbf{A}_1 + [\Psi_1(\dot{\gamma}) + \Psi_2(\dot{\gamma})]\,\mathbf{A}_1^2 - \frac{1}{2}\Psi_1(\dot{\gamma})\,\mathbf{A}_2 \tag{2.31}$$

$$\Psi_i = \frac{N_i}{(\dot{\gamma})^2}, \qquad i = 1,2$$

N_i are the first and the second normal stress differences and $\dot{\gamma}$ represents the shear rate. This equation is called the CEF (Criminale–Ericksen–Filbey) equation. It is exact not only for simple shearing flows, but for all flows where $\forall \mathbf{A}_k = 0$ for $k \geq 3$ and has found many applications. The second-order fluid model (2.30) is the same as the CEF equation (2.31) in structure except that second-order fluid model has constant coefficients. The latter applies to any flow that is *slow and slowly varying* and the former to any *steady shearing* flow even if the flow as a whole is not viscometric.

As sound as the foundation of the theory may be, unfortunately no unique way of specifying the memory functionals, the kernels in a series of multiple integral approximation to the response functional, has been found as yet and hence predictions are very difficult if not impossible. The restrictions imposed on the constitutive functionals by thermodynamics and invariance leave enough room to allow an enormous choice of memory functionals rendering the method of little value for practical applications. In addition, it should be noted that in any approximation to the response functional be it order fluids of the integral type, order fluids of the differential type or differential fluids of grade n the number of material constants rises rapidly with increasing order n of the approximation posing seemingly insurmountable experimental difficulties in determining these constants. Siginer [116] notes that although it is next to impossible to determine a large number of parameters from a single experiment of rheometry, it may be possible to determine the parameters sequentially from a series of rheometrical experiments at the lower orders, and outlines a series of experiments involving free surface rheometry and pulsating flow experiments in tubes to determine the material constants of the integral fluid of order three with fading memory. In spite of these difficulties, integral fluid of order two with fading memory has been successfully used to describe some quasi-unsteady flows such as rod climbing on oscillating rods and to determine the material constants in the second functional derivative of the fluid of order two, Beavers [112], and the fluid of order three with the third functional derivative in (2.23) or (2.24) has been used to investigate two nearly viscometric flows, non-linear effects of tube wall vibration on the Poiseuille flow of viscoelastic fluids, Siginer [110], in torsional oscillations of a layered medium of immiscible viscoelastic liquids driven by an oscillating rod, Siginer [117], and pulsating flow of rheologically complex fluids in round tubes, Siginer [111], Siginer and Letelier [118] and Letelier et al. [119].

2.5 Constitutive Equation Formulations Consistent with Thermodynamics

Recent developments related to compatibility of *rate type* CEs with thermodynamics and the restrictions on the material constants for compliance with the second law are interesting. For incompressible fluids, a rearrangement of (2.30) with p representing the pressure and truncation at the third order reads as:

$$\mathbf{T} = -p\mathbf{1} + \mu\mathbf{A}_1 + \alpha_1\mathbf{A}_2 + \alpha_2\mathbf{A}_1^2 + \beta_1\mathbf{A}_3 + \beta_2(\mathbf{A}_1\mathbf{A}_2 + \mathbf{A}_2\mathbf{A}_1) + \beta_3(tr\mathbf{A}_1^2)\mathbf{A}_1$$

$$\mathbf{L} = \text{grad}\,\mathbf{u}, \quad 2\mathbf{D} = \mathbf{L} + \mathbf{L}^{\mathrm{T}}$$

$$\mathbf{A}_1 = 2\mathbf{D}, \quad \mathbf{A}_{n+1} = \mathbf{A}_n + \mathbf{L}^{\mathrm{T}}\mathbf{A}_n + \mathbf{A}_n\mathbf{L}$$

where p is the constitutively indeterminate part of the total stress, **1** is the unit tensor, μ is the zero shear viscosity, and α_1, α_2 together with β_1 , β_2 , β_3 are material constants. Considering this equation to be an exact model in its own right in the sense described by Fosdick and Rajagopal [120], not an approximation places restrictions on the material coefficients via thermodynamical considerations unlike the case of the retardation approximation. The Clausius–Duhem inequality is required to hold and the Helmholtz free energy is constrained to be a minimum when the fluid is locally at rest, in particular $\beta_1 = \beta_2 = 0$ and $\mu \geq 0$, $\alpha_1 \geq 0$, $\beta_3 \geq 0$, $|\alpha_1 + \alpha_2| \leq \sqrt{24\mu\beta_3}$ due to these thermodynamical requirements reducing the thermodynamics compliant form to:

$$\mathbf{T} = -p\mathbf{1} + \mu\mathbf{A}_1 + \alpha_1\mathbf{A}_2 + \alpha_2\mathbf{A}_1^2 + \beta_3(tr\mathbf{A}_1^2)\mathbf{A}_1$$

Trouble is available experimental data indicates that $\alpha_1 < 0$ for the working fluids in the experiments, which poses a dilemma as it contradicts the thermodynamically dictated restrictions worked out by Fosdick and Rajagopal [120]. One may then conjecture that the fluids in the experiments are not fluids of grade three or they abide by a more encompassing constitutive equation of which the fluid of grade three is the truncated form. There has been no definitive resolution to this day to these conflicting experimental and analytical results. However, some light has been shed on the issue by the seminal paper of Müller and Wilmanski [121]. Thermodynamics places the restrictions $\mu > 0$, $\alpha_1 > 0$, $\alpha_1 + \alpha_2 = 0$ on the material constants of the second-order fluid taken as an exact model in its own right for compatibility with the second law, Dunn and Fosdick [122]. However experiments indicate that $\mu > 0$, $\alpha_1 < 0$, $\alpha_1 + \alpha_2 \neq 0$. Dunn and Fosdick [122] also showed that disregarding thermodynamics and assuming that $\alpha_1 < 0$ leads to instabilities and in quite arbitrary flows instability and boundedness are unavoidable. That the instabilities persist even if $\mu > 0$, $\alpha_1 < 0$, $\alpha_1 + \alpha_2 \neq 0$ is assumed was shown by Fosdick and Rajagopal [123]. In addition, Joseph [114] has shown that for the second-order fluid, and indeed for fluids of arbitrary higher order, the rest state is unstable. Müller and Wilmanski [121] rederive the second-order fluid model within the framework of the *extended irreversible thermodynamics* introduced by Müller [124] and find that α_1 may be negative. They also find that there are no instabilities and the speeds of the shear waves determined by the viscosity and the normal stress coefficients are finite. The same holds for the family of all *rate type* equations of which the second-order fluid is part of. Extended thermodynamics puts different restrictions on the material parameters of *rate type* equations. The idea behind *extended irreversible thermodynamics* is to take the stress and the heat flux as *independent* variables. Each one of them is governed by a balance equation. As balance equations are not conservation equations, they include fluxes. The immediate consequence is that the classical constitutive equations, the Navier–Stokes theory and the Fourier's law, are *no longer constitutive equations* in their own right but rather arise at the first order of the Maxwellian iteration process, Müller [124], Müller and Wilmanski [121]. Following the lead of Müller and Wilmanski the Reiner–Rivlin and Rivlin–Ericksen second-order fluids are derived by Lebon and Cloot [125] in the framework of *extended irreversible thermodynamics* and the resulting equations with the attached thermodynamics restrictions imposed by extended thermodynamics was used to analytically solve the problem of the Marangoni convection in a thin horizontal layer of a non-Newtonian fluid subjected to a temperature gradient in microgravity. Although it is not directly related to non-Newtonian fluid mechanics, we note that Depireux and Lebon [126] studied non-Fickian diffusion in a two-component mixture at uniform temperature in the framework of extended irreversible thermodynamics. The idea here is to elevate the dissipative diffusion flux to the status of independent variable and derive an evolution equation for the diffusion flux to show that in the linear regime the classical Fick's laws are recovered. The results in the non-linear regime lead to non-Fickian diffusion laws different from the well-known Fick's laws.

A thermodynamically consistent framework to derive CEs to model motions of viscoelastic fluids with arbitrarily large strains and strain rates using an internal parameter based on ideas with their origin in solid mechanics was introduced and developed by Leonov in a series of publications culminating in his 1987 review paper, Leonov [33]. For instance, the classical solid mechanics idea of storing elastic energy is interesting because some important phenomena, such as the dependence of viscosity on rate of shear and the normal stress effects such as the die-swell and rod climbing may be connected with the ability of the fluid to store elastic energy temporarily. The framework of deriving CEs with a single relaxation mode based on the recoverable strain tensor, an internal parameter, which arises from the formalism of irreversible thermodynamics together with the kinematics of the flow, is presented in detail with examples in [33]. The rheology of polymeric fluids with narrow molecular weight distribution (MWD) can be described by a single relaxation mode in particular at relatively high shear rates as the shear rate is increased although the description will not agree with experimental data at low shear rates as the shear rate is further reduced. Several relaxation modes are needed if the MWD of the polymeric system is rather wide. Leonov demonstrates that a rheological single mode theory with a single tensorial internal parameter is powerful enough to describe all the viscoelastic phenomena in the standard test cases of simple shear and uniaxial elongation. The approach advocated by Leonov [33] based on quasi-linear irreversible thermodynamics with the recoverable strain tensor considered as an internal parameter is not as general as the "fading memory" approach with memory functionals, but nevertheless contains considerable arbitrariness due to the many unknown material functions that are only weakly restricted by the dissipative inequalities imposed by thermodynamical compliance considerations. But the firm thermodynamic basis makes it preferable to the traditional rheological approach introduced by Oldroyd. In addition, it does not lead to non-evolutionary equations, it can be used in non-isothermal and compressible flows, and it can yield information about the recoverable strain tensor, which is often measured experimentally.

In all fairness as innovative and original that it may be the pioneering thermodynamically consistent theories with internal parameters with their origin in solid mechanics advocated by Leonov to derive CEs for viscoelastic fluids was built on the work of Kluitenberg [127–129] who was the first to propose the kinematic idea of dividing the total infinitesimal strain tensor in viscoelastic media into recoverable and irreversible parts. In his own words Leonov's contribution was "a thermodynamic description of the viscoelastic behavior of polymer-like liquids for the general case of arbitrary and finite recoverable strains" (p. 7 in [33]). That of course does not detract in any way from the achievement of Leonov in devising the first manageable thermodynamic theory of the viscoelastic fluid behavior.

Mattos' [34] work is another example of thermodynamically consistent approach to devise a constitutive theory with an internal parameter for rate type materials within the framework of the thermodynamics of irreversible processes. Dissipative phenomena can be accounted for only if the second law of thermodynamics is part of the methodology. The procedure described by Mattos to obtain constitutive relations permits a generalization of classical rheological models of Oldroyd [27], Maxwell [30], and Jeffreys [32] by including microstructural effects. Helmholtz free energy is assumed to be an isotropic and differentiable function of a finite set of independent and objective variables that can also be used to model the interaction of microstructure with the macrostructure; thus, for instance Mattos' method allows internal spin unlike the framework developed by Leonov. It should be noted that internal spin can be included in Leonov's methodology, but so far this has not been done. The original theory presented by Leonov [33] does not include internal spin. The variables in the Helmholtz free energy in Mattos' theory are the absolute temperature, the density, and a measure of strain. The choice of a particular objective time derivative and of two thermodynamic potentials is sufficient to develop a methodology to define a complete set of constitutive equations. Temperature plays an important role in complex flows of viscoelastic fluids in many industrial applications, such as injection molding of polymers. Mattos' theory allows an adequate modeling of thermomechanical couplings when dissipation due to changes of the material structure must also be taken into account.

The rate of dissipation d based on the second law of thermodynamics distinguishes between admissible ($d \geq 0$) and inadmissible processes ($d < 0$). If the rate of dissipation d is always equal to zero, the admissible process is reversible. In terms of the Helmholtz free energy per unit mass ψ, Cauchy stress $\mathbf{T}$, the stretching tensor $\mathbf{D}$, internal energy per unit mass e, absolute temperature θ, total entropy per unit mass s, and the heat flux vector $\mathbf{q}$, the local form of the second law of thermodynamics, the Clausius–Duhem inequality reads as:

$$d = \mathbf{T} \bullet \mathbf{D} - \rho\left(\dot{\psi} + s\,\dot{\theta}\right) + \mathbf{q} \bullet \mathbf{g} \geq 0 \tag{2.32}$$

$$\psi = e - \theta s, \quad \mathbf{g} = -\mathrm{grad}\,(\log\theta)$$

where $(\dot{\bullet})$ indicates the material time derivative of the entity between the parentheses. The class of fluids considered is that for which the Helmholtz free energy ψ can be expressed as an isotropic and differentiable function of the absolute temperature θ, of the density ρ, of a measure of strain $\boldsymbol{\varepsilon}$ whose objective time derivative is the stretching tensor $\mathbf{D}\left[\overset{\nabla}{\boldsymbol{\varepsilon}}\right] = \mathbf{D}$ [see (2.33) for the definition of the objective time derivative $\left(\overset{\nabla}{\bullet}\right)$ used], and a number of other independent objective auxiliary variables that may be introduced into the constitutive model theory to describe various facets of the interaction of the microstructure of the material with its macrostructure such as a scalar variable for instance to account for the proportion of broken connections between structural units in a macromolecule. In Mattos' theory, these variables are related to dissipative mechanisms, or in other words to irreversible changes of the microstructure. For simplicity, Mattos [34] considers only one such additional independent objective variable $\mathbf{A}$:

$$\psi = \psi(\rho, \mathbf{D}, \mathbf{A}, \theta)$$

Since the material time derivative of an objective tensorial quantity is not necessarily objective, the use of special time derivatives in rate type constitutive equations is required to ensure objectivity. The choice of a particular derivative is tantamount to a constitutive assumption. A number of choices can be made for the objective time derivative $\left(\overset{\nabla}{\bullet}\right)$ any one of which appears as the material derivative of $(\bullet)$ to an observer in a frame of reference attached to the particle and rotating with it at an angular velocity equal to the instantaneous value of the spin of the particle. Mattos [34] assumes an objective time derivative $\left(\overset{\nabla}{\bullet}\right)$ of the following type in terms of the asymmetric part $\mathbf{W}$ of the velocity gradient tensor $\mathbf{L}$, or spin tensor as it is also called sometimes, and a skew-symmetric second- order tensor $\mathbf{W}^R$, which is associated with the micro-motions of the material structure, the relative rotation of the material structure with respect to the continuum.

$$\overset{\nabla}{(\bullet)} = \dot{(\bullet)} + (\bullet)\left(\mathbf{W} - \mathbf{W}^R\right) - \left(\mathbf{W} - \mathbf{W}^R\right)(\bullet) \tag{2.33}$$

If $\mathbf{W}^R = 0$ (2.33) collapses onto the Jaumann objective derivative. $(\mathbf{W} - \mathbf{W}^R)$ is the spin of the particle associated with the independent movement of a unit vector

$\mathbf{n}$ attached to the material point $\dot{\mathbf{n}} = \left(\mathbf{W} - \mathbf{W}^R\right)\mathbf{n}$. The constitutive assumption that governs the rate of energy dissipation d is expressed as a function of the stretching tensor $\mathbf{D}$ and the objective derivative (2.33) of the auxiliary variable $\mathbf{A}$ that relates the microstructure to macrostructure, $d_m = \left(\mathbf{D}, \overset{\triangledown}{\mathbf{A}}\right) \ni d_m = (\mathbf{0}, \mathbf{0}) = 0$.

The coupling of the changes at the microstructure level with the macrostructure can be achieved not only by the choice of the thermodynamic potentials ψ and ϕ and their arguments but also by the choice of the relative spin $\mathbf{W}^R$. Besides the possible couplings caused by products of the variables ρ, $\boldsymbol{\varepsilon}$, $\mathbf{A}$, and θ in the free energy potential, the objective derivative chosen may induce a secondary coupling between different physical mechanisms depending on the expression adopted for $\mathbf{W}^R = \boldsymbol{f}(\mathbf{T}, \mathbf{D}, \mathbf{A})$ the full expression for which is quite complicated. Mattos [34] chooses to use the first term $\mathbf{W}^R = \mathrm{C}\,(\mathbf{AD} - \mathbf{DA})$ [C is a constant] only in the full expression for $\mathbf{W}^R = \boldsymbol{f}(\mathbf{T}, \mathbf{D}, \mathbf{A})$ on the premise that is good enough for an adequate phenomenological description of the fluid's behavior. The following form (assumption) is introduced for $d_m = \left(\mathbf{D}, \overset{\triangledown}{\mathbf{A}}\right)$:

$$d_m = \frac{\partial \phi}{\partial \mathbf{D}} \bullet \mathbf{D} + \frac{\partial \phi}{\partial \overset{\triangledown}{\mathbf{A}}} \bullet \overset{\triangledown}{\mathbf{A}} \geq 0 \quad \forall \left(\mathbf{D}, \overset{\triangledown}{\mathbf{A}}\right) \tag{2.34}$$

where ϕ is a differentiable function of $\mathbf{D}$ and $\overset{\triangledown}{\mathbf{A}}$ such that $\phi(\mathbf{0}, \mathbf{0}) = 0$. Equation (2.34) implies that the total energy dissipation due to mechanical effects is the sum of the rate of energy dissipation due to viscous phenomena [the first term on the RHS of (2.34)] and dissipation due to changes in the microstructure of the material [the second term on the RHS of (2.34)]. The inequality of the second law of thermodynamics is automatically satisfied in any process if (2.34) holds, no matter the nature of the external stimuli, the boundary and the initial conditions. The potentials ψ and ϕ are not required to be convex functions. The convexity of ϕ is not necessary but is sufficient for (2.34) to be valid.

Using (2.32) and (2.33) and (2.34), it can be shown that the following constitutive relations always hold if incompressibility $\mathrm{tr}\mathbf{D} = 0$ is imposed:

$$s = -\frac{\partial \psi}{\partial \theta}, \quad \rho \frac{\partial \psi}{\partial \mathbf{A}} - \frac{\partial \phi}{\partial \overset{\triangledown}{\mathbf{A}}} = 0, \quad \mathbf{T} = -p\mathbf{1} + \rho \frac{\partial \psi}{\partial \boldsymbol{\varepsilon}} + \frac{\partial \phi}{\partial \mathbf{D}} \tag{2.35}$$

The choice of the expression for the relative spin $\mathbf{W}^R$ and for the potentials ϕ and ψ completely characterizes a given fluid in isothermal flows. The addition of the Fourier's law together with (2.35) forms a complete set of thermodynamics compliant constitutive equations for fluids undergoing non-isothermal flows. For compressible flows $\mathrm{tr}\mathbf{D} \neq 0$ $(2.35)_3$ is replaced with

$$\mathbf{T} = -\rho^2 \frac{\partial \psi}{\partial \rho}\mathbf{1} + \rho \frac{\partial \psi}{\partial \boldsymbol{\varepsilon}} + \frac{\partial \phi}{\partial \mathbf{D}} \tag{2.36}$$

where the first term on the RHS represents the thermodynamic pressure. The second term on the RHS of $(2.35)_3$ and (2.36) represents the reversible or elastic part and the third term the irreversible part due to viscous dissipation. A more general theory can be constructed if ϕ is conceived of to be a function of $\phi = \phi\left(\mathbf{D}, \overset{\triangledown}{\mathbf{A}}, \rho, \boldsymbol{\varepsilon}, \mathbf{A}, \theta\right)$. Mattos shows how the general theory yields the generalized Newtonian fluid, the Bingham fluid, and the viscoelastic Maxwell model. If ψ is a differentiable isotropic function of θ and $\mathbf{A}$

$$\psi(\mathbf{A}, \theta) = \psi_m(\mathbf{A}) + \psi_t(\theta)$$

and the potential ϕ is a differentiable, isotropic function of $\mathbf{D}$ and $\overset{\triangledown}{\mathbf{A}}$ not identically equal to zero, one gets from (2.35):

$$\mathbf{T} = -p\mathbf{1} + \frac{\partial \phi}{\partial \mathbf{D}}, \quad \rho \frac{\partial \psi_m}{\partial \mathbf{D}} - \frac{\partial \phi}{\partial \overset{\triangledown}{\mathbf{A}}} = 0$$

which can be reduced to a model with shear-rate-dependent viscosity. If on the other hand,

$$\psi(\rho, \boldsymbol{\varepsilon}, \mathbf{A}, \theta) = \psi_m(\rho, \boldsymbol{\varepsilon}, \mathbf{A}) + \psi_t(\theta) = \frac{1}{\rho}\hat{\psi}(\boldsymbol{\varepsilon}, \mathbf{A}) + \psi_t(\theta)$$

with $\hat{\psi}$ a differentiable function of ($\boldsymbol{\varepsilon}$, $\mathbf{A}$) one gets from (2.35),

$$\mathbf{T} = -p\mathbf{1} + \frac{\partial \hat{\psi}_m}{\partial \boldsymbol{\varepsilon}} + \frac{\partial \phi}{\partial \mathbf{D}}$$

$$\frac{\partial \hat{\psi}_m}{\partial \mathbf{A}} - \frac{\partial \phi}{\partial \overset{\triangledown}{\mathbf{A}}} = 0$$

which can be reduced to the generalized Maxwell CE if $\hat{\psi}$ and ϕ have the following forms:

$$\hat{\psi}_m(\boldsymbol{\varepsilon}, \mathbf{A}) = \frac{1}{2}\lambda\ (\boldsymbol{\varepsilon} - \mathbf{A}) \bullet (\boldsymbol{\varepsilon} - \mathbf{A})$$

$$\phi\left(\overset{\triangledown}{\mathbf{A}}\right) = \frac{1}{2}\kappa\ \overset{\triangledown}{\mathbf{A}} \bullet \overset{\triangledown}{\mathbf{A}}$$

where λ and κ are constitutive constants.

2.5.1 Maximization of the Rate of Dissipation

A thermodynamic framework that leads to a rational methodology for developing constitutive relations for the viscoelastic response of materials with instantaneous elasticity was worked out by Rajagopal and Srinivasa [35]. The methodology provides a unified basis to study a wide class of material response such as traditional plasticity, twinning, solid-to-solid phase transition, multinetwork theory, and includes, as special subcases, classical elasticity, classical linearly viscous fluid, and viscoelasticity. The viscoelastic response is determined by a stored energy function that characterizes the elastic response from the "natural configuration" of the material and a rate of dissipation function that describes the rate of dissipation due to viscous effects. It is noted that the material body may possess a single, many, or an infinite number of natural configurations. For instance, the classical elastic solid has a single natural configuration, whereas the traditional plastic material and the classical linear viscous fluid have an infinite number of natural configurations. So, the theory is actually based on the evolution of the natural configuration of the material. The evolution of the natural configuration driven by external stimuli is determined by a thermodynamic criterion, which shapes the response function of the material to maximize the entropy production. The evolution of the natural configuration is determined by the rate of dissipation, or to be more precise, the maximization of the rate of dissipation. In a purely mechanical context, the response of the material is characterized by constitutively prescribing the stored energy (or Helmholtz potential) and the rate of dissipation functions. Since in a closed system the entropy increases to achieve its maximum equilibrium value, the quickest way in which the maximum could be reached is by maximizing the rate of dissipation.

In constructing CEs compatible with the thermodynamics of irreversible processes, the usual procedure is to require that the second law of thermodynamics in its general form is satisfied that is the rate of entropy production is non-negative for all processes. Taking for instance the case of viscous fluids with implicit CEs, that is fluids with CEs of the type $\mathcal{F}(\mathbf{T}, \mathbf{D}) = 0$ or of the more general type:

$$\mathcal{F}\left(\mathbf{T}, \overset{\triangledown}{\mathbf{T}}, \ldots, \overset{\overset{(n)}{\triangledown}}{\mathbf{T}}, \mathbf{D}, \overset{\triangledown}{\mathbf{D}}, \ldots, \overset{\overset{(n)}{\triangledown}}{\mathbf{D}}\right) = 0$$

where $\overset{\overset{(n)}{\triangledown}}{(\bullet)}$ $\mathbf{T}$ and $\mathbf{D}$ stand for the n Oldroyd derivatives, total stress and the rate of deformation tensors) when the viscosity is pressure-dependent and the CE is no longer an explicit relationship between the Cauchy stress and the kinematical variables, the usual procedure is to assume a CE for the total (Cauchy) stress $\mathbf{T}$ in terms of the rate of deformation $\mathbf{D}$ and absolute temperature θ, assume that Fourier's law holds for the heat flux vector q and use the non-negativity of the rate of entropy production ξ for all processes to obtain certain restrictions on constitutive equations. However, if a certain rate of entropy production is desired, additional assumptions are required to choose a specific CE from the class of admissible entropy production functions (response functions) that will produce the specific entropy production rate and will define by the same token the rate of change of the state variables of the material. Clearly such a choice would lead to a non-negative entropy production in terms of the state variables.

The idea that among all competing constitutive assumptions for the selection of an appropriate entropy producing process the one which maximizes the rate of entropy production is the right choice goes back to Ziegler [130]. Rajagopal and Srinivasa in a series of publications starting in 1998 [131] first related to solid mechanics and later extended to cover fluids [35] further clarified and developed this principle and applied it to material CEs in entropy producing processes. In a later publication, Rajagopal and Srinivasa [132] show that linear phenomenological evolution laws that satisfy the Onsager relations such as Fourier's law of heat conduction, Fick's law, Darcy's law, Newton's law of viscosity, and others all corroborate the assumption that quadratic forms for the rate of entropy production lead to linear phenomenological relations that satisfy the Onsager relations. For other forms of entropy production that are not quadratic for which the Onsager relations and related theorems cannot be applied they discuss how the ideas of Onsager can be generalized to include non-linear phenomenological laws and they outline a procedure to obtain non-linear laws. In addition to further developing the maximum entropy production approach to selecting a CE to describe the behavior of the material, they show that this approach is not in contradiction of the well-known Onsager [133] and Prigogine [134] principles that lead to the minimization of the rate of entropy production. They clarify that the Onsager and Prigogine principles refer to totally different circumstances. Maximization of the rate of entropy production leads to a constitutive choice amongst a competing class of constitutive relations in Rajagopal and Srinivasa's theory resulting in a non-negative rate of entropy production, which can be viewed as a Lyapunov function. This Lyapunov function reaches a minimum as the body tends towards equilibrium in time. It is this latter minimum that is referred to as the "Onsager's principle."

Rajagopal and Srinivasa were interested in [35] in modeling viscoelastic fluids, which exhibit instantaneous elastic response. The central idea at the very foundation of their theory is that the body possesses numerous natural configurations. The response of the material is "elastic" from these natural configurations, and the rate of dissipation determines how these natural configurations evolve. Rajagopal and Srinivasa [35] postulate that the evolution of the natural configuration is described by maximizing the rate of dissipation function. They recognize that this requirement is not necessarily a fundamental principle and that there may be other prescriptions to define the evolution of the natural configuration. For instance, different forms of the stored energy function and the rate of dissipation function lead to different models, which describe different types of responses. They show that the choice of a neo-Hookean elastic response for the stored energy function from the current natural configuration of the material and a maximized rate of dissipation function that is quadratic in the stretching tensor associated with the current configuration lead to Maxwell-like models, and another choice for the rate of dissipation function leads to an Oldroyd-B type of model, which under further restrictions of small elastic strains reduces to the Oldroyd-B model.

It is well known that the introduction of the frame-invariant time derivatives leads to non-linear models. Thus whereas the one-dimensional Maxwell model is linear, the frame-invariant higher dimensional model is non-linear. The theory of Rajagopal and Srinivasa [35] is not based on a generalization from one dimension to three dimensions. Consequently there is no need to choose a particular frame-invariant rate a priori. The rates that appear in the constitutive equation are dictated by the choice of the stored energy function and the rate of dissipation function. The three-dimensional counterpart of the well-known one-dimensional Maxwell model, with the mechanical analog of a spring and a dashpot in series, is rigorously derived in [35]. It is also shown that if the displacement gradients are small, this non-linear, three-dimensional version of the Maxwell model collapses onto the classical upper-convected Maxwell model.

In a more recent 2009 paper Karra and Rajagopal [135], following up the earlier work of Rajagopal and Srinivasa [35], extend the thermodynamic framework developed in [35] for rate-type models for viscoelastic fluids *with* instantaneous elasticity to rate-type models for viscoelastic fluids *without* instantaneous elasticity. They start from the premise that the response from the natural configuration to the current configuration is like that of a generalized Kelvin–Voigt solid. When the external load is removed, the body moves back to the natural configuration with some "relaxation time" which is greater than the intrinsic time t_m. If this relaxation time is set to a value less than the intrinsic time t_m, the class of models that can be generated using the framework developed in [35] is obtained. Using the thermodynamic framework developed, they derive constitutive relations which in one dimension reduce to a dashpot and a Kelvin–Voigt element (a spring and a dashpot in parallel) in series whose viscosities are stretch dependent.

Given the field equations with the divergence operator ($\nabla\bullet$) taken with respect to the current configuration κ_t, Fig. 2.6:

$$\nabla\bullet\mathbf{u}=0,\quad \rho\frac{D\mathbf{u}}{Dt}=\nabla\bullet\mathbf{T}^{\mathrm{T}}+\rho\mathbf{f},\quad \mathbf{T}^{\mathrm{T}}=\mathbf{T} \tag{2.37}$$

where $\mathbf{u}$, $\mathbf{T}$, $\mathbf{f}$, and ρ stand for the velocity field, the Cauchy stress tensor, the body force field, and the density, and the local form of balance of energy as:

$$\rho\frac{De}{Dt}=\mathbf{T}\bullet\mathbf{L}-\nabla\bullet\mathbf{q}+\rho r \tag{2.38}$$

where e, $\mathbf{L}$, $\mathbf{q}$, and r represent the specific internal energy per unit mass, the velocity gradient tensor, the heat flux vector, and the specific radiant energy, respectively, [the gradient in the material derivative in (2.37) and (2.38) is taken with respect to the current configuration κ_t, Fig. 2.6], and the second law of thermodynamics in the form of the energy dissipation equation as:

$$\mathbf{T}\bullet\mathbf{D}-\rho\dot{\psi}=\rho\theta\zeta=\xi\geq 0 \tag{2.39}$$

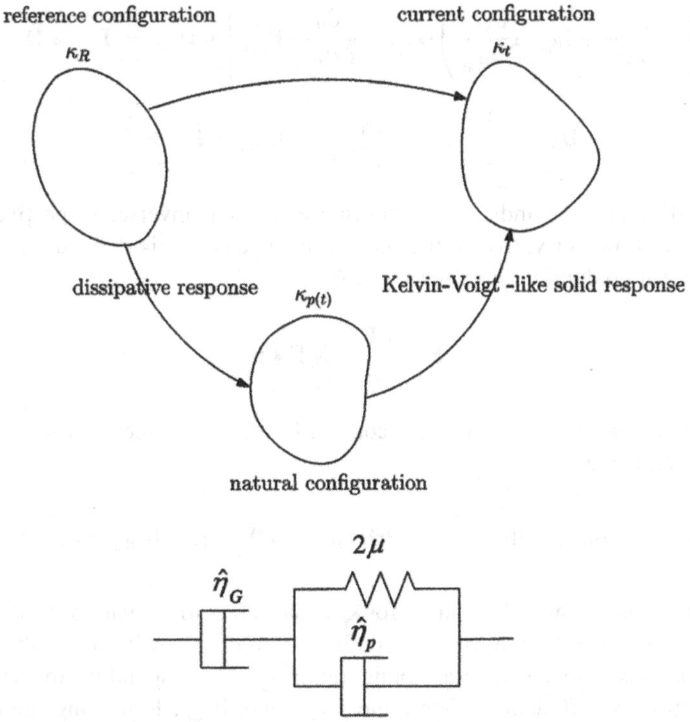

Fig. 2.6 The natural configuration $\kappa_{p(t)}$ corresponding to the current configuration κ_t and the relevant mappings from the tangent spaces of the same material point in κ_R, κ_t, and $\kappa_{p(t)}$. The response from the natural configuration $\kappa_{p(t)}$ is like a Kelvin–Voigt solid and the response of $\kappa_{p(t)}$ from the reference configuration κ_R is purely dissipative. The corresponding one-dimensional spring dashpot analogy consists in a dashpot in series with a Kelvin–Voigt element as shown (Adapted from Karra and Rajagopal [135] with permission)

where $\mathbf{T}, \mathbf{D}$ are the Cauchy stress and the rate of deformation tensor, respectively, ψ is the specific Helmholtz free energy, ζ is the rate of entropy production, and ξ is the rate of dissipation with θ representing the absolute temperature. The specific Helmholtz free energy is chosen to be a function of the first and second invariants of the Cauchy–Green left stretch tensor $\mathbf{B}_{p(t)}$ as:

$$\psi = \psi\left(\mathbf{B}_{p(t)}\right) = \hat{\psi}\left(\mathrm{I}_{\mathbf{B}_{p(t)}},\ \mathrm{II}_{\mathbf{B}_{p(t)}}\right),\ \mathbf{B}_{p(t)} = \mathbf{F}_{p(t)}\mathbf{F}_{p(t)}^{\mathrm{T}} \tag{2.40}$$

The exact form cannot be determined given that infinity of functions could pass through any finite number of given experimental points. $\mathbf{F}_{p(t)}$ defines the mapping from the tangent space at a material point in the natural configuration $\kappa_{p(t)}$ to the tangent space at the same material point in the current configuration κ_t, see Fig. 2.6. The following comes out of (2.40):

$$\dot{\psi} = 2\left[\left(\frac{\partial \hat{\psi}}{\partial \mathrm{I}_{\mathbf{B}_{p(t)}}} + \mathrm{I}_{\mathbf{B}_{p(t)}} \frac{\partial \hat{\psi}}{\partial \mathrm{II}_{\mathbf{B}_{p(t)}}}\right)\mathbf{B}_{p(t)} - \frac{\partial \hat{\psi}}{\partial \mathrm{II}_{\mathbf{B}_{p(t)}}}\mathbf{B}^2_{p(t)}\right] \bullet \mathbf{D}_{p(t)} = \mathbf{T}_{p(t)} \bullet \mathbf{D}_{p(t)} \quad (2.41)$$

$$\mathbf{D}_{p(t)} = \frac{1}{2}\left(\mathbf{L}_{p(t)} + \mathbf{L}^{\mathrm{T}}_{p(t)}\right), \quad \mathbf{L}_{p(t)} = \dot{\mathbf{F}}_{p(t)}\mathbf{F}^{-1}_{p(t)}$$

Where $(\bullet)^{\mathrm{T}}$ and $(\bullet)^{-1}$ indicate the transpose and the inverse, respectively, and the material time derivative of the second-order tensor $\mathbf{F}$ is defined with respect to the current configuration κ_t, see Fig. 2.6,

$$\dot{\mathbf{F}} = \frac{\partial \mathbf{F}}{\partial t} + \nabla \mathbf{F} \bullet \mathbf{u}$$

and the principal invariants of the second-order Cauchy–Green left stretch tensor $\mathbf{B}_{p(t)}$ are defined as:

$$\mathrm{I}_{\mathbf{B}_{p(t)}} = tr\mathbf{B}_{p(t)}, \quad \mathrm{II}_{\mathbf{B}_{p(t)}} = \frac{1}{2}\left[\left(tr\mathbf{B}_{p(t)}\right)^2 - tr\mathbf{B}^2_{p(t)}\right], \quad \mathrm{III}_{\mathbf{B}_{p(t)}} = \det\mathbf{B}_{p(t)}$$

The index $p(t)$ is an abbreviation for $\kappa_{p(t)}$ and refers to the natural configuration $\kappa_{p(t)}$ corresponding to the current configuration κ_t. Knowledge of the current configuration κ_t and the natural configuration $\kappa_{p(t)}$ corresponding to the current configuration is sufficient to determine $\mathbf{F}_{\kappa_{\mathbf{p(t)}}}$ and $\mathbf{B}_{\kappa_{\mathbf{p(t)}}}$. Removing the external stimuli present in the current configuration κ_t responsible for the deformation will take the body back to the natural configuration $\kappa_{p(t)}$. The natural configuration and hence $\mathbf{F}_{\kappa_{\mathbf{p(t)}}}$ is determined by instantaneous elastic unloading for viscoelastic fluids endowed with instantaneous elastic response. For viscoelastic fluids without instantaneous elastic response, the natural configuration is obtained by removing the external stimuli in a way consistent with the class of thermodynamic processes that are allowable.

The rate of dissipation is a function of the stretching tensor $\mathbf{D}_{p(t)}$ and Cauchy–Green left stretch tensor $\mathbf{B}_{p(t)}$ between the $\kappa_{p(t)}$ to κ_t, the stretching tensor $\mathbf{D}_G$ between κ_R and $\kappa_{p(t)}$:

$$\xi = \xi\left(\mathbf{B}_{p(t)}, \mathbf{D}_{p(t)}, \mathbf{D}_G\right) \quad (2.42)$$

$\mathbf{G}$ is defined as the mapping that takes κ_R into $\kappa_{p(t)}$,

$$\mathbf{G} = \mathbf{F}_{\kappa_R \to \kappa_{p(t)}} = \mathbf{F}^{-1}_{\kappa_{p(t)}}\mathbf{F}_{\kappa_R}, \quad \mathbf{L}_G = \dot{\mathbf{G}}\mathbf{G}^{-1}, \quad \mathbf{D}_G = \frac{1}{2}\left(\mathbf{L}_G + \mathbf{L}^{\mathrm{T}}_G\right)$$

The body dissipates energy during its motion from $\kappa_{p(t)}$ to κ_t as well as during its motion from κ_R to $\kappa_{p(t)}$, the former because of the healing of the polymer networks and the latter due to the sliding of polymer chains over one another. The rate of dissipation ξ is maximized using (2.39) and the isochoric motion assumption $\mathrm{tr}\mathbf{D}_{p(t)} = \mathrm{tr}\mathbf{D}_G = 0$ as constraints [tr(·) is the trace operator for the second-order tensors] to derive the CEs flowing out of the rate of dissipation assumption (2.42). The maximization process which uses the method of "Lagrange multipliers" starts by substituting (2.41) into (2.39),

$$\mathbf{T} \bullet \mathbf{D} - \rho \mathbf{T}_{p(t)} \bullet \mathbf{D}_{p(t)} = \xi \ \left(\mathbf{B}_{p(t)}, \mathbf{D}_{p(t)}, \mathbf{D}_G\right)$$

Details can be found in [135]. To illustrate the power of the method developed, specific choices (2.43) and (2.44) are made for the Helmholtz potential and the rate of dissipation, respectively, as:

$$\psi = \hat{\psi}\left(\mathrm{I}_{\mathbf{B}_{p(t)}},\ \mathrm{II}_{\mathbf{B}_{p(t)}}\right) = \frac{\mu}{2\rho}\left(\mathrm{I}_{\mathbf{B}_{p(t)}} - 3\right) \tag{2.43}$$

$$\xi\left(\mathbf{B}_{p(t)}, \mathbf{D}_{p(t)}, \mathbf{D}_G\right) = \eta_p \mathbf{D}_{p(t)} \bullet \mathbf{B}_{p(t)} \mathbf{D}_{p(t)} + \eta_G \mathbf{D}_G \bullet \mathbf{B}_{p(t)} \mathbf{D}_G \tag{2.44}$$

The stored energy chosen here is that for a neo-Hookean material with μ its elastic modulus, whereas the rate of dissipation is similar to that of a "mixture" of two Newtonian-like fluids (in the sense that the dissipation is quadratic in the symmetric part of the velocity gradient), whose dissipation also depends on the stretch (specifically the stretch from the natural configuration to the current configuration), with viscosities η_G and η_p. The former term on the right-hand side of (2.44) is due to the dissipation during the motion from κ_R to $\kappa_{p(t)}$ and the latter term is due to dissipation during the motion from $\kappa_{p(t)}$ to κ_t. Note that with the above choices for ψ and ξ, as the body moves from the $\kappa_{p(t)}$ to κ_t, there is both storage (like a neo-Hookean solid) and dissipation (like a Newtonian-like fluid) of energy simultaneously, and hence $\kappa_{p(t)}$ evolves like the natural configuration of a Kelvin–Voigt-like solid with respect to κ_t.

With the specific forms for the Helmholtz potential and the rate of dissipation (2.43) and (2.44) adopted the rate of dissipation is maximized with the constraint that the difference between the stress power and the rate of change of Helmholtz potential is equal to the rate of dissipation and any other constraint that may be applicable such as incompressibility. The class of models developed exhibit fluid-like characteristics incapable of instantaneous elastic response when none of the material moduli that appear in the model are not zero. Maxwell-like and Kelvin–Voigt-like viscoelastic materials spawn out of this class when some material moduli are assigned special values. A specific model which stores energy like a neo-Hookean solid is derived with a rate of dissipation which depends on the stretching tensor associated with the natural configuration and the stretching tensor between the natural configuration and the current configuration, see Fig. 2.6. The model reduces to either the Maxwell-like fluid or the Kelvin–Voigt-like solid under certain restrictions on the material parameters.

It is interesting to note that instead of (2.39) one could equally well have chosen a rate of dissipation without any stretch dependence and quadratic in the symmetric part of the velocity gradient:

$$\xi(\mathbf{D}_{p(t)}, \mathbf{D}_G) = \eta_p \mathbf{D}_{p(t)} \bullet \mathbf{D}_{p(t)} + \eta_G \mathbf{D}_G \bullet \mathbf{D}_G$$

to end up after the maximization process with a variant of the CEs, which result from the adoption of (2.43), Karra and Rajagopal [135].

2.5.2 Burgers Equation

One-dimensional Burgers equation (2.2) for linear viscoelasticity was presented in Sect. 2.3.1. Burgers equation is increasingly important to seemingly disconnected branches of science and engineering as many uses in practice from polymeric fluids to the description of the deformation of geological materials are common. The generalization of the one- dimensional Burgers equation to three dimensions on a thermodynamically compatible basis was worked out recently by Quintanilla and Rajagopal [43, 44] using the thermodynamic framework developed by Rajagopal and Srinivasa [35] to systematically generate CEs for rate-type viscoelastic fluids with instantaneous elasticity. They show that Cauchy stress in a generalized Burgers fluid is given by:

$$\mathbf{T} = -p\mathbf{1} + \mathbf{S}$$
$$\mathbf{S} + \lambda_1 \overset{\triangledown}{\mathbf{S}} + \lambda_2 \overset{\triangledown\triangledown}{\mathbf{S}} = \eta_1 \mathbf{D} + \eta_2 \overset{\triangledown}{\mathbf{D}} + \eta_3 \overset{\triangledown\triangledown}{\mathbf{D}}$$
$$\overset{\triangledown}{\mathbf{A}} = \frac{D\mathbf{A}}{Dt} - \mathbf{LA} - \mathbf{AL}^{\mathrm{T}}, \quad \mathbf{A} = (\mathbf{S}, \mathbf{D}) \tag{2.45}$$

where **T**, **D**, and **S** represent the total stress, the symmetric part of the velocity gradient **L**, and the extra-stress tensor, respectively. $(-p\mathbf{1})$ is the indeterminate part of the stress due to the constraint of incompressibility, and the notation $(\overset{\triangledown\triangledown}{\bullet})$indicated the application of the Oldroyd derivative operator twice. For isochoric motions $tr\mathbf{D} = 0$, substituting (2.45) into the linear momentum equation with $\mathbf{u}$ and f representing the velocity and body force fields:

$$\rho \frac{D\mathbf{u}}{Dt} = \nabla \bullet \mathbf{T} + \rho f$$

and linearizing they obtain the following system:

$$\rho \frac{\partial \tilde{\mathbf{u}}}{\partial t} = -\nabla q + \Delta \hat{\mathbf{u}} + \rho \boldsymbol{b} \tag{2.46}$$

$$\tilde{\mathbf{u}} = \mathbf{u} + \lambda_1 \frac{\partial \mathbf{u}}{\partial t} + \lambda_1 \frac{\partial^2 \mathbf{u}}{\partial t^2} \tag{2.47a}$$

$$\hat{\mathbf{u}} = \eta_1 \mathbf{u} + \eta_2 \frac{\partial \mathbf{u}}{\partial t} + \eta_3 \frac{\partial^2 \mathbf{u}}{\partial t^2} \tag{2.47b}$$

They note that when $\eta_3 = 0$, $\eta_3 = 0 = \lambda_2$, and $\eta_2 = \eta_3 = \lambda_2 = 0$, the linear form of the Burgers fluid, the linear form of the Oldroyd-B fluid, and the linear version of the Maxwell fluid, respectively, are recovered. And of course, the Navier–Stokes representation for constitutively linear fluids is obtained when $\lambda_1 = \eta_2 = \eta_3 = \lambda_2 = 0$. The boundary-initial-value problem defined by the system (2.46) and (2.47a, 2.47b) and the boundary-initial conditions:

$$\mathbf{u}(\mathbf{x}, t)|_{\partial\Sigma} = \mathbf{0}, \quad \mathbf{u}(\mathbf{x}, 0) = \mathbf{u}^0, \quad \frac{\partial \mathbf{u}}{\partial t}(\mathbf{x}, 0) = \mathbf{u}_1^0, \quad \frac{\partial^2 \mathbf{u}}{\partial t^2}(\mathbf{x}, 0) = \mathbf{u}_2^0, \quad \mathbf{x} \in \Sigma \tag{2.48}$$

is studied in a three-dimensional bounded domain Σ with smooth boundary $\partial\Sigma$. They show in [43] that the stability condition for Burgers fluids ($\eta_3 = 0$) is $\lambda_1\eta_2 > \eta_1\lambda_2$. They prove in [44] continuous dependence on the initial data and uniqueness of unsteady solutions for generalized Burgers fluids ($\eta_3 \neq 0$) in bounded three-dimensional domains whenever conditions $\lambda_i > 0$, $\eta_i > 0$ are met, and the exponential decay of solutions of the homogeneous problem (2.46), (2.47a, 2.47b), and (2.48) ($b_i = 0$) in three-dimensional bounded domains Σ with smooth boundary $\partial\Sigma$ when in addition to $\lambda_i > 0$, $\eta_i > 0$ the following conditions hold $\lambda_1\, \eta_3 > \lambda_2\, \eta_2$, $\eta_3 > \lambda_2\, \eta_1$, $\eta_2 > \lambda_1\, \eta_1$. Next they ask the question what happens if $\lambda_i > 0$ and $\eta_i > 0$ are still valid but $\lambda_1\, \eta_3 > \lambda_2\, \eta_2, \eta_3 > \lambda_2\, \eta_1, \eta_2 > \lambda_1\, \eta_1$ do not hold, and they prove that the solutions to uniaxial shearing flows are unstable if $\lambda_i > 0$ and $\eta_i > 0$ hold but, $\eta_3 + \lambda_1\eta_2 > \lambda_2\eta_1$, $\eta_3 + \lambda_1\eta_2 \leq \lambda_2\eta_1$, $[\lambda_2\eta_1 - (\eta_3 + \lambda_1\eta_2)]^2 < 4\eta_2\eta_3\lambda_1$ do not hold. As the condition of stability in the case of the Burgers fluids ($\eta_3 = 0$) is $\lambda_1\eta_2 > \lambda_2\eta_1$ the class of parameters for the generalized Burgers fluids ($\eta_3 \neq 0$) where stability is expected is larger than the class for the Burgers fluid.

A major step in devising thermodynamically consistent CEs is to recognize that several different sets of stored energy and rate of dissipation function can lead to the same three-dimensional non-linear CE. Karra and Rajagopal [136] consider four different sets of stored energy and rate of dissipation functions to obtain four seemingly different three-dimensional sets of CEs for the stress, each one of which with equal claim to the status of three-dimensional generalization of the Burgers model as all four can be made to collapse on the one-dimensional model developed by Burgers [39, 40]. The choice of two scalar functions for the stored energy and the rate of dissipation leads to a CE for the stress, a tensor with six scalar components. Historically, the development of many of the one-dimensional CEs to describe the response of viscoelastic materials were based on analogies with mechanical systems of springs (to store energy), and dashpots (to dissipate energy/produce entropy). Karra and Rajagopal [136] demonstrate within the context of these mechanical spring–dashpot systems how the same CE for the stress (Burgers equation) can be obtained by choosing different stored energy and rate of entropy production functions each one of which would correspond to a different network of springs and dashpots with the same response.

A body capable of instantaneous elastic response that exists in a configuration κ_t under the action of external stimuli, on the removal of the external stimuli could return to a configuration $\kappa_{p(t)}$ referred to as *a* natural configuration corresponding to the configuration κ_t. However, more than one natural configuration could be associated with the configuration κ_t based on the rate of removal of the external stimuli. Instantaneous removal and removal with a finite speed will imply return to different natural configurations. The natural configuration that is accessed depends on the process class allowed. Karra and Rajagopal [136] assume that the natural configuration is the one due to instantaneous unloading, the body responding in an instantaneous elastic manner. They note that even within the context of instantaneous elastic unloading, it might be possible that the body may go to different natural configurations $\kappa_{p_i(t)}$, $i = 1, \ldots, n$. That is because a point is a mathematical creation that does not exist, and in reality what is modeled is a sufficiently small neighborhood of a point in the body. Energy can be stored and dissipated in different ways by the material enclosed in this neighborhood. Various arrangements of springs and dashpots can lead to the same net storage of energy of the springs and the dissipation by the dashpots. In other words, the chunk of material in the neighborhood of the point can respond in an identical manner for different ways in which the springs and dashpots are put together. Karra and Rajagopal [136] work with two natural configurations to incorporate in the CEs to be derived two relaxation times possessed by Burgers-like fluid bodies.

The second law of thermodynamics merely requires that the entropy production be non-negative. The requirement of maximization of the rate of entropy advocated by Rajagopal et al. [35, 41–44, 131, 132, 135, 136] narrows the range of choices one can make from the class of rate of entropy production functions. In the following, the process for the derivation of the first set of three-dimensional CEs detailed in [136] will be outlined.

Let κ_R denote the undeformed reference configuration of the body. It is assumed that the body has two natural configurations to which it can be instantaneously elastically unloaded. This implies two different mechanisms of storing energy, which would correspond within one-dimensional mechanical analog to two different springs. The body can get from the reference configuration to the two evolving natural configurations denoted by $\kappa_{p_i(t)}$, $i = 1, 2$ (Fig. 2.7) via two dissipative responses. Let F_i, $i = 1, 2, 3$ denote the gradients of the motion from κ_R to $\kappa_{p_1(t)}$, $\kappa_{p_1(t)}$ to $\kappa_{p_2(t)}$, and $\kappa_{p_2(t)}$ to κ_t, respectively.

Defining the left Cauchy–Green stretch tensors $\mathbf{B}_i$ and the velocity gradients $\mathbf{L}_i$ with their corresponding symmetric parts $\mathbf{D}_i$

$$\mathbf{B}_i = \mathbf{F}_i\mathbf{F}_i^{\mathrm{T}}\,,\ \mathbf{L}_i = \dot{\mathbf{F}}_i\mathbf{F}_i^{-1},\ \mathbf{D}_i = \tfrac{1}{2}\left(\mathbf{L}_i + \mathbf{L}_i^{\mathrm{T}}\right), \quad i = 1, 2, 3,$$

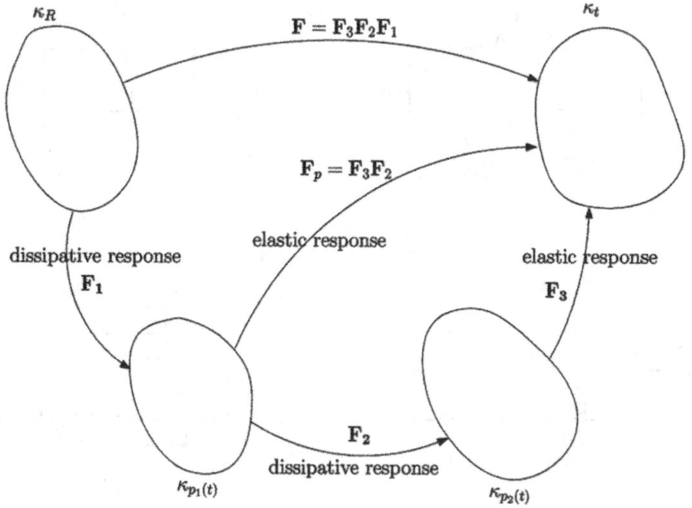

Fig. 2.7 Illustration of the natural configurations for the model used: κ_R is the reference configuration, κ_t denotes the current configuration; $\kappa_{p_1(t)}$ and $\kappa_{p_2(t)}$ denote the two evolving natural configurations. The body dissipates energy like a viscous fluid as it moves from κ_R to $\kappa_{p_1(t)}$ and from $\kappa_{p_1(t)}$ to $\kappa_{p_2(t)}$. The body stores energy during its motion from $\kappa_{p_2(t)}$ to κ_t, and $\kappa_{p_1(t)}$ to κ_t (Adapted from Karra and Rajagopal [136] with permission)

the specific stored energy ψ and the rate of dissipation ξ are assumed to be of the following form, with the gradient of the motion from $\kappa_{p_1(t)}$ to κ_t denoted by $\mathbf{F}_p$, given that the instantaneous elastic responses from $\kappa_{p_1(t)}$ and $\kappa_{p_2(t)}$ are isotropic:

$$\psi = \psi(\mathbf{B}_3, \mathbf{B}_p), \quad \xi = \xi(\mathbf{D}_1, \mathbf{D}_2) \tag{2.49}$$

$$\mathbf{B}_p = \mathbf{F}_p\mathbf{F}_p^{\mathrm{T}}, \quad \mathbf{F}_p = \mathbf{F}_3\mathbf{F}_2, \quad \mathbf{L}_p = \dot{\mathbf{F}}_p\mathbf{F}_p^{-1}$$

In virtue of incompressibility of the body, the specific forms are further narrowed down to:

$$\psi = \psi(\mathbf{B}_3, \mathbf{B}_p) = \frac{\mu_3}{2\rho}(\mathbf{1} \bullet \mathbf{B}_3 - 3) + \frac{\mu_p}{2\rho}(\mathbf{1} \bullet \mathbf{B}_p - 3) \tag{2.50}$$

$$\xi = \xi(\mathbf{D}_1, \mathbf{D}_2) = \eta'_1\ \mathbf{D}_1 \bullet \mathbf{D}_1 + \eta'_2\ \mathbf{D}_2 \bullet \mathbf{D}_2 \tag{2.51}$$

In this, μ_3, and μ_p are the elastic moduli and η'_1 and η'_2 are the viscosities which account for the energy storage and energy dissipation, respectively. The assumptions (2.50) and (2.51) mean that the body possesses instantaneous elastic response from the two evolving natural configurations $\kappa_{p_1(t)}$ and $\kappa_{p_2(t)}$ to the current configuration κ_t (Fig. 2.7); the body stores energy like a neo-Hookean solid during its motion, from $\kappa_{p_1(t)}$ to κ_t, and from $\kappa_{p_2(t)}$ to κ_t. In addition, the response is linear viscous fluid like, as

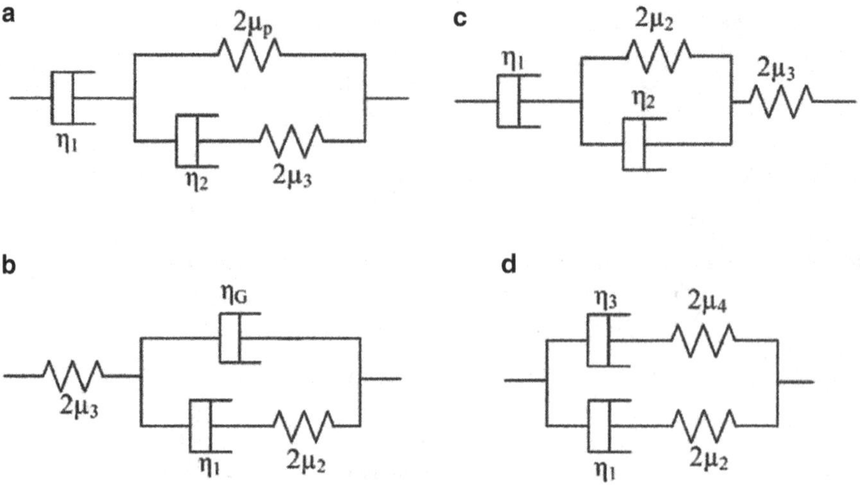

Fig. 2.8 Various spring-dashpot arrangements which reduce to the one-dimensional Burgers' fluid model (Adapted from Karra and Rajagopal [136] with permission)

the body moves from κ_R to $\kappa_{p_1(t)}$ and from one natural configuration $\kappa_{p_1(t)}$ to the other $\kappa_{p_2(t)}$. The assumptions (2.50) and (2.51) lead to the CEs (2.52) and (2.53).

$$\begin{gathered} \mathbf{T} = -p\mathbf{1} + \mu_3\ \mathbf{B}_3 + \mu_p\mathbf{B}_p, \\ (\mu_3\ \mathbf{B}_3)^2 = \frac{1}{3}\mu_3^2 tr(\mathbf{B}_3)\mathbf{B}_3 - \frac{\eta_2}{2}\mu_3\ \overset{\triangledown}{\mathbf{B}}{}_3^{p} \end{gathered} \tag{2.52}$$

$$\frac{1}{2}\mu_3\mu_p\left(\mathbf{B}_p\mathbf{B}_3 + \mathbf{B}_3\mathbf{B}_p\right) + \mu_p\mathbf{B}_p{}^2 = \frac{1}{3}\left[tr(\mu_3\ \mathbf{B}_3) + tr\left(\mu_p\mathbf{B}_p\right)\right]\ \mu_p\mathbf{B}_p - \frac{\eta_1}{2}\ \mu_p\ \overset{\triangledown}{\mathbf{B}}_p \tag{2.53}$$

The notation $\overset{\triangledown}{(\bullet)}{}^{p}$ and $\overset{\triangledown}{(\bullet)}$, indicates the Oldroyd derivatives, the former when the natural configuration $\kappa_{p_1(t)}$ is made the reference configuration κ_R, and the viscosities η_i, $i = 1, 2$ are redefined versions of the original viscosities $\eta_i^{'}$, $i = 1, 2$ in (2.51). They proceed to show that (2.52) and (2.53) can be reduced in one dimension to the one- dimensional Burgers equation (2.54):

$$\sigma + \left(\frac{\eta_2}{2\mu_p} + \frac{\eta_2}{2\mu_3} + \frac{\eta_1}{2\mu_p}\right)\ \dot{\sigma} + \frac{\eta_1\eta_2}{4\mu_p\mu_3}\ \ddot{\sigma} = \eta_1\ \dot{\varepsilon} + \frac{\eta_1\eta_2}{2\mu_p}\ \left(1 + \frac{\mu_p}{\mu_3}\right)\ \ddot{\varepsilon} \tag{2.54}$$

corresponding to the spring–dashpot mechanical analog (a) in Fig. 2.8 [comparison of (2.54) with (2.2) is invited]. σ and ε denote the stress and the linearized strain *in one dimension*, respectively, and the notation $(\dot{\bullet})$ and $(\ddot{\bullet})$ indicates first and second derivatives with respect to time.

Karra and Rajagopal [136] proceed to show that the specific stored energy ψ and the rate of dissipation ξ can be chosen in three more different ways, similar but different than (2.50) and (2.51), each set of which yields a different set of three-dimensional CEs of the same type as (2.52) and (2.53), all of which can be reduced in one dimension to the one-dimensional Burgers equation (2.2) corresponding, respectively, to the mechanical analogs (b–d) shown in Fig. 2.8.

2.5.3 Minimum Free Energy and Maximum Recoverable Work

Many studies done on the free energy of materials with memory both solids and fluids have shown that the free energy is not necessarily uniquely defined. There may be many free energies associated with any given state of these material bodies. Free energy expressions form a bounded and convex set with a minimum and a maximum element. The determination of explicit forms for the free energy has become an interesting problem, the study of which started in the 1960s has continued to this day spawning several expressions that have been proposed for *viscoelastic solids*, Breuer and Onat [137], Day [138], Graffi [139, 140], Morro and Vianello [141], Graffi and Fabrizio [142], Fabrizio and Morro [143], Fabrizio et al. [144, 145], Del Piero and Deseri [146], Deseri et al. [147], Fabrizio and Golden [148], and Deseri et al. [149]. Recent papers by Amendola [150, 151] and Amendola and Fabrizio [152] addressed the problem of determining a general closed form explicit expression for the isothermal minimum free energy of *isotropic, incompressible, and linearly viscoelastic fluids* with fading memory together with the maximum recoverable work related to the minimum free energy. A general closed expression is given for the isothermal minimum free energy of a linearized incompressible viscoelastic fluid, whose constitutive equation is expressed by a linear functional of the history of strain by Amendola [150] who also in a follow-up work adapted to linearized incompressible viscoelastic fluids some expressions for the free energy previously proposed and studied for viscoelastic solids, Amendola [151]. The maximum recoverable work related to the minimum free energy of the system given in terms of Fourier-transformed quantities for *incompressible linearized viscoelastic fluids* was investigated by Amendola and Fabrizio [152] and corresponding expressions were derived. That the minimum free energy is related to the maximum work obtainable from a given material at a given state $\sigma(t)$ *starting from an initial state* $\sigma_i(t)$ was already shown in the 1960s and early 1970s, Breuer and Onat [137, 153] and Day [138]. The work of Amendola is built on the method and the procedure introduced by Golden [154] and Gentili [155], respectively, in the scalar case for *linear viscoelastic solid materials*. They work with the linearized form for small strains of the general CE (2.12) or (2.13). The particle which is at $\boldsymbol{x}$ at the present time t in an Eulerian frame was at $\mathbf{r}(\mathbf{x}, t')$ at time $t' > t$, and the extra-stress tensor is a functional of the past history of the strain tensor $\mathbf{C}$. Let the relative displacement $\mathbf{u}$ be defined as $\mathbf{r} - \mathbf{x} = \mathbf{u}(\mathbf{S}, \mathbf{x})$, where $\mathbf{S}$ represents the extra-stress, and assume that $\frac{\partial \mathbf{u}}{\partial \mathbf{x}} << 1$ which implies that higher powers of $\frac{\partial \mathbf{u}}{\partial \mathbf{x}}$ can be neglected in deriving an explicit CE from (2.12) or (2.13). Then the following holds:

$$\mathbf{C} = \mathbf{F}^{\mathrm{T}}\mathbf{F} \to 2\mathbf{C} = \frac{\partial \mathbf{u}}{\partial \mathbf{x}} + \left(\frac{\partial \mathbf{u}}{\partial \mathbf{x}}\right)^{\mathrm{T}} + O\left(\left|\frac{\partial \mathbf{u}}{\partial \mathbf{x}}\right|^{2}\right) \to 2\mathbf{E} = \frac{\partial \mathbf{u}}{\partial \mathbf{x}} + \left(\frac{\partial \mathbf{u}}{\partial \mathbf{x}}\right)^{\mathrm{T}} = \nabla \mathbf{u} + (\nabla \mathbf{u})^{\mathrm{T}}$$

where $\mathbf{E}$ is the small strain tensor. Assuming now that the extra-stress $\mathbf{S}$ depends linearly on $\mathbf{E}$ for small enough strains, the following linearized form of (2.12) is obtained:

$$\mathbf{T}(\mathbf{x}, t) = -p(\mathbf{x}, t)\,\mathbf{1} + 2\int_0^\infty \zeta(\mathbf{x}, s)\ \left[\mathbf{E}^t(\mathbf{x}, s) - \mathbf{E}(\mathbf{x}, t)\right] ds \qquad (2.55)$$

where $\mathbf{E}(\mathbf{x},t)$ is the instantaneous value of the strain at time t and $\mathbf{E}^t(\mathbf{x},s) = \mathbf{E}(\mathbf{x}, t - s)$ $\forall\, s \in (0, +\infty) = \mathbf{R}^{++}$ is the past history of the strain. The material function $\zeta(\mathbf{x},s)$ is such that the shear relaxation modulus $G(s)$ given by:

$$G(s) \in L^1([0, \infty)) \to G(s) = -\int_s^\infty \zeta(\tau)\, d\tau, \quad \lim_{s \to +\infty} G(s) = 0, \quad G(s) > 0$$

must satisfy restrictions imposed on the constitutive equation (2.55) by thermodynamic principles. The material defined by (2.55) is a simple material in the sense defined by Coleman and Owen [156]. The thermodynamic restrictions on the constitutive equation (2.55) have been derived by Fabrizio and Lazzari [157] who proved the following theorem: *The constitutive equation* (2.55) *for linear viscoelastic fluids is compatible with the Second Law of Thermodynamics* ***if and only if*** *for every relaxation modulus G* (*s*):

$$G(s) \in L^1\ ([0, \infty)) \colon \int_0^\infty G(s) ds \neq 0$$

the following inequality holds:

$$\int_0^\infty G(s)\ \cos(\omega s)\, ds > 0 \quad \forall \omega \in \mathbf{R}$$

where $\mathbf{R}$ represents the set of real numbers. The constitutive equation (2.55) characterizes the behavior of a simple fluid and as such its properties can be described in terms of states $\sigma(t)$ and processes P, Noll [158]. The stress-power w reads as:

$$w[\sigma(t), P(t)] = \mathbf{T}(t) \bullet \mathbf{L}(t) = \mathbf{T}(t) \bullet \mathbf{D}(t), \quad \mathbf{T}^{\mathrm{T}} = \mathbf{T}, \quad 2\mathbf{D} = \mathbf{L} + \mathbf{L}^{\mathrm{T}}$$

The strain rate tensor by virtue of the linear approximation collapses onto the time derivative of the small strain tensor **E**, that is $\mathbf{D}(t) = \dot{\mathbf{E}}(t)$. The state $\sigma(t)$ of a system is given by the *relative strain history* $\mathbf{E}_r^t(s)$ expressed in terms of the instantaneous value $\mathbf{E}(\mathbf{x},t)$ of the strain and the past history of the strain $\mathbf{E}^t(\mathbf{x},s)$ as:

$$\sigma(t) = \mathbf{E}_r^t(s) = [\mathbf{E}^t(\mathbf{x}, s) - \mathbf{E}(\mathbf{x}, t)] \quad \forall s \in (0, \infty) = \mathbf{R}^{++}$$

The work W done on the material on a path γ up to time t going from an initial state $\sigma_i(t)$ to a state $\sigma(t) = \sigma(\sigma_i,P) = \mathbf{E}_r^t$ by means of an admissible process P is given by:

$$W(\sigma, P) = \int_{\gamma} \mathbf{T}(s) \bullet \dot{\mathbf{E}}(s)\, ds = \int_{-\infty}^{t} \mathbf{T}(s) \bullet \mathbf{D}(s)\, ds$$

$$= 2 \int_{-\infty}^{t} \left(\int_{0}^{+\infty} \zeta(s)\, \mathbf{E}_r^{\tau}(s)\, ds \right) \bullet \dot{\mathbf{E}}(\tau) d\tau$$

If the process is *isothermal*, the second law of Thermodynamics collapses onto the dissipation principle, the *Clausius–Duhem* inequality, which stipulates that the work done during any cycle (σ,P) must satisfy:

$$W(\sigma, P) = \oint \mathbf{T}(t) \bullet \mathbf{D}(t)\, dt \ \geq 0$$

where the equality sign applies only to reversible cases, Fabrizio and Morro [159]. *The minimum free energy* $\psi_m[\sigma(t)]$ *coincides with the maximum recoverable work* $W_R[\ \sigma(t)\ ]$ *that is the maximum work that can be extracted from a given state* $\sigma(t)$ *of the fluid* $\psi_m[\sigma(t)] = W_R[\sigma(t)]$. Amendola [151] has shown that the maximum free energy ψ_M in the case of materials of type (2.55) is given by:

$$\psi_M[\sigma(t)] = \int_0^t \mathbf{T}\left[\mathbf{E}_r^t(s)\right] \bullet \dot{\mathbf{E}}(s)\, ds$$

and that any other free energy $\psi[\ \sigma(t)]$ corresponding to a different process would satisfy the inequality:

$$\psi[\sigma(t)] \leq \int_0^t \mathbf{T}(s) \bullet \dot{\mathbf{E}}(s)\, ds \leftrightarrow \dot{\psi} \leq \mathbf{T}\left[\mathbf{E}_r^t(s)\right] \bullet \dot{\mathbf{E}}(s)$$

This can be written as an equality by introducing a non-negative dissipation function $D(t)$:

$$\dot{\psi} + D(t) = \mathbf{T}\left[\mathbf{E}_r^t(t)\right] \bullet \dot{\mathbf{E}}(t) \tag{2.56}$$

Amendola considers free energy expressions based on *Graffi–Volterra functionals* already introduced for viscoelastic solids, Graffi [139, 140] and Volterra [160]. The corresponding functional for linearized viscoelastic fluids of type (2.55) is given by:

$$\psi_G(t) = -\int_0^{\infty} \zeta(s)\, \mathbf{E}_r^t(s) \bullet \mathbf{E}_r^t(s)\, ds, \quad \zeta(s) \leq 0, \quad \frac{d\zeta(s)}{ds} \geq 0 \quad \forall s \in \mathbf{R}^+$$

The internal dissipation $D_G(\,t)$ a non-negative entity for all admissible histories is given as:

$$D_G(t) = \int_0^{+\infty} \frac{d\varsigma(s)}{ds} \mathbf{E}_r^t(s) \bullet \mathbf{E}_r^t(s)\, ds \geq 0$$

Day's [138] free energy worked out for viscoelastic solids can also be adapted to the case of linearized viscoelastic fluids of type (2.55) as:

$$\psi_D(t) = \frac{1}{G(0)} \left[\int_0^{+\infty} \varsigma(s)\ \mathbf{E}_r^t(s)\ ds \right]^2$$

There are other forms of free energy formulations based on different functionals such as the most general representation for free energy, the *Breuer–Onat functional* previously studied in the context of viscoelastic solids as well [137, 153].

$$\psi(t) = \int_0^{\infty} \int_0^{\infty} G_{12}(s,u)\, \mathbf{E}_r^t(u) \bullet \mathbf{E}_r^t(s)\, du\, ds \tag{2.57}$$

$$G_{12}(s,u) = \frac{\partial^2}{\partial s \partial u} G(s,u), \quad G_{12}(+\infty,u) = G_{12}(s,+\infty) = 0 \tag{2.58}$$

The last two conditions are required because the integral in (2.57) must exist for *finite relative histories* $\mathbf{E}_r^t$.

$$\begin{aligned} &G(s,u) = \int_s^{+\infty} \int_u^{+\infty} G_{12}(s',\ u')\ ds' du', \quad G(+\infty,u) = G(s,+\infty) = 0, \\ &G(0,s) = G(s,0) = G(s) \end{aligned} \tag{2.59}$$

Integration by parts yields:

$$\psi(t) = \int_0^\infty \int_0^\infty G(s,u)\, \dot{\mathbf{E}}^t(u) \bullet \dot{\mathbf{E}}^t(s)\, du\, ds \tag{2.60}$$

A dissipation function which satisfies (2.56) is also derived by evaluating the time derivative of (2.60) as:

$$\dot{\psi} = \mathbf{T}(t) \bullet \dot{\mathbf{E}}(t) + \int_0^{+\infty}\int_0^{+\infty} [G_1(s,u) + G_2(s,u)]\dot{\mathbf{E}}^t(u) \bullet \dot{\mathbf{E}}^t(s)\, du\, ds \tag{2.61}$$
$$G_1(s, +\infty) = G_2(+\infty, u) = 0, \quad G_1(s,0) = G_2(0,s) = \zeta(s)$$

The *Breuer–Onat* internal dissipation is obtained from (2.56) and (2.61) as:

$$\begin{aligned} D_m(t) &= -\int_0^{+\infty}\int_0^{+\infty} \mathcal{L}_{12}(s,u)\, \mathbf{E}_r^t(u) \bullet \mathbf{E}_r^t(s)\, du\, ds \\ &= -\int_0^\infty \int_0^\infty \mathcal{L}(s,u)\, \dot{\mathbf{E}}^t(u) \bullet \dot{\mathbf{E}}^t(s)\, du\, ds \ \geq \ 0 \end{aligned} \tag{2.62}$$

$$\mathcal{L}_{12}(s,u) = \frac{\partial^2}{\partial s \partial u}\mathcal{L}(s,u), \quad \mathcal{L}(s,u) = G_1(s,u) + G_2(s,u) \tag{2.63}$$

The free energy of Golden [154] originally developed for linear viscoelastic solids was derived for linear viscoelastic fluids of type (2.55) by Amendola [151]. It reads as:

$$\psi_m(t) = \frac{1}{\pi}\int_{-\infty}^{+\infty} \left| \mathbf{q}^t_{(-)}\ (\omega)\right|^2 d\omega \tag{2.64}$$

$$\begin{aligned} \mathbf{q}^t_{(-)}\ (\omega) &= -\lim_{z\to\omega^-} \frac{1}{2\pi i}\int_{-\infty}^{+\infty} \frac{i\omega' G_{(-)}(\omega')\ \mathbf{E}^t_{r+}(\omega')}{(\omega' - z)}\, d\omega', \\ \omega^- &= \lim_{\alpha\to 0^+} (\omega - i\alpha), \qquad z \in \mathbf{C} \end{aligned} \tag{2.65}$$

C represents the set of $\forall$ complex numbers. The subscript ($\pm$) is used to indicate that the corresponding function of ω, a function of $z \in \mathbf{C}$, has zeros and singularities only for $z \in \mathbf{C}^{\pm}$, and $(\bullet)_+$ represents the half-range Fourier transform $(\bullet)$: $R \to R^n$:

$$(\bullet)_+(\omega) = \int_0^{+\infty} (\bullet)(s) e^{-i\omega s} ds,$$
$$(\bullet)(\omega) = \int_{-\infty}^{+\infty} (\bullet)(s) e^{-i\omega s} ds = (\bullet)_+(\omega) + (\bullet)_-(\omega), \quad \forall \omega \in R \tag{2.66}$$

The Fourier transform of the strain history $\mathbf{E}_r^t$ is denoted by $\mathbf{E}_{r+}^t$. Details of these complex computational developments can be found in [151].

A relaxation modulus for linear viscoelastic solids defined as a discrete spectrum of a sum of exponentials was considered by Golden [154] who derived the corresponding minimum free energy with related internal dissipation in the frequency domain. His method and example has been applied by Amendola and Fabrizio [152] to linearized viscoelastic fluids of type (2.55), which will be summarized in some detail below as an example.

$$G(t) = \begin{cases} \mu_i e^{-\alpha_i t} & \forall t \geq 0 \\ 0 & \forall t < 0 \end{cases}, \quad \mu(0) = \sum_{i=1}^{n} \mu_i > 0, \quad \alpha_1 < \alpha_2 < \ldots. < \alpha_{n-1} < \alpha_n,$$
$$\alpha_i > 0, \quad \mu_i > 0$$

To calculate the free energy of the material from (2.64) first the half range Fourier (cos) transform $G_c(\omega)$ of $G(t)$ is computed as:

$$G_c(\omega) = \sum_{i=1}^{n} \frac{\alpha_i \mu_i}{\alpha_i^2 + \omega^2}$$

Defining a new function $\mathcal{K}(\omega)$ with no zeros at any $\omega \in R$ or at infinity, in terms of Fourier half range transforms (2.66) such that $\mathcal{K}(\omega) = (1+\omega^2) G_c(\omega) = \mathcal{K}_{(+)}(\omega)\ \mathcal{K}_{(-)}(\omega)$. It follows that $G_c(\omega) = G_{(+)}(\omega)\, G_{(-)}(\omega)$ and $G_{(\pm)}(\omega) = \mathcal{K}_{(\pm)}(\omega)\ (1 \pm i\omega)^{-1}$,

$$\mathcal{K}(\omega) = \sum_{i=1}^{n} \alpha_i \mu_i \frac{1+\omega^2}{\alpha_i^2 + \omega^2} \to \mathcal{K}(\omega) = \mathcal{K}_\infty \prod_{m=1}^{n} \left(\frac{\gamma_m^2 + \omega^2}{\alpha_m^2 + \omega^2} \right),$$
$$\mathcal{K}_\infty = \lim_{\omega \to \infty} \mathcal{K}(\omega) = \sum_{i=1}^{n} \alpha_i \mu_i > 0 \tag{2.67}$$

where $\gamma_1^2 = 1$ and γ_j^2, $j = 1, 2, \ldots, n$ denote the simple zeros of the analytic function$f(z) = \mathcal{K}(\omega)$ with $z = -\omega^2$. At most only one of the values γ_j^2 can coincide with $\gamma_1^2 = 1$, which then becomes a zero of multiplicity 2. The factorization $(2.67)_2$ yields together with $G_{(\pm)}(\omega) = \mathcal{K}_{(\pm)}(\omega)\ (1 \pm i\omega)^{-1}$:

$$G_{(-)}(\omega) = i k_\infty \omega^{-1} \prod_{m=1}^{n} \left(\frac{\omega + i\delta_m}{\omega + i\alpha_m} \right) = i k_\infty \omega^{-1} \left(1 + i \sum_{r=1}^{n} \frac{A_r}{\omega + i\alpha_r} \right), \quad k_\infty = \sqrt{\mathcal{K}_\infty}$$

$$A_r = (\delta_r - \alpha_r) \prod_{i=1, i \neq r}^{n} \left(\frac{\delta_i - \alpha_r}{\alpha_i - \alpha_r} \right), \quad r = 1, 2, \ldots, n; \quad \delta_j, \ j = 2, 3, \ldots, n$$

where $\delta_1 = \gamma_o = 0$; $\delta_j = \gamma_j$, $j = 2, 3, \ldots, n$. When $n \neq 1$

$$\mathbf{q}^t_{(-)}(\omega) = k_\infty \sum_{m=1}^{n} A_m \frac{\mathbf{E}^t_{r+}(-i\alpha_m)}{\omega + i\alpha_m}$$

$$\mathbf{E}^t_{r+}(-i\alpha_m) = \int_0^{+\infty} \mathbf{E}^t_r(s) \ \mathrm{e}^{-\alpha_m s} ds$$

$$\psi_m(t) = 2K_\infty \sum_{i,j=1}^{n} \frac{A_i A_j}{\alpha_i + \alpha_j} \left\{ \int_0^\infty \int_0^\infty \mathrm{e}^{-(\alpha_i s_1 + \alpha_j s_2)} \mathbf{E}^t_r(s_1) \bullet \mathbf{E}^t_r(s_2) ds_1 ds_2 \right\}$$

For the particular case when $n = 1$, the expression for the free energy is obtained through (2.65) and (2.64) as:

$$\mathbf{q}^t_{(-)}(\omega) = k_\infty A_1 \frac{\mathbf{E}^t_{r+}(-i\alpha_1)}{\omega + i\alpha_1} \rightarrow \psi_m(t) = \mu_1 \alpha_1^2 \left[\int_0^\infty \mathrm{e}^{-\alpha_1 s} \mathbf{E}^t_r(s) \ ds \right]^2$$

The expression for the related internal dissipation when $n = 1$ reads as:

$$D_m(t) = 2\mu_1 \alpha_1^3 \left[\int_0^\infty \mathrm{e}^{-\alpha_1 s} \mathbf{E}^t_r(s) \ ds \right]^2$$

Computing *Breuer–Onat* free energy Amendola and Fabrizio [152] find from $(2.59)_1$ and (2.57) as:

$$G_{12}(s, u) = \frac{2}{\mathcal{K}_\infty} \sum_{i,j=1}^{n} \frac{\alpha_i^2 \alpha_j^2 \mu_i \mu_j}{(\alpha_i + \alpha_j) \ B_i B_j} \mathrm{e}^{-(\alpha_i s_1 + \alpha_j s_2)} > 0$$

$$\psi_m(t) = \frac{2}{\mathcal{K}_\infty} \sum_{i,j=1}^{n} \frac{\alpha_i^3 \alpha_j^3 \mu_i \mu_j}{(\alpha_i + \alpha_j) B_i B_j} \left\{ \int_0^\infty \int_0^\infty \mathrm{e}^{-(\alpha_i s_1 + \alpha_j s_2)} \ \mathbf{E}^t_r(s_1) \bullet \mathbf{E}^t_r(s_2) \ ds_1 ds_2 \right\}$$

$$B_r = (\delta_r + \alpha_r) \prod_{l=1, l\neq r}^{n} \left[\frac{\delta_l + \alpha_r}{\alpha_l + \alpha_r}\right] > 0, \quad r = 1, 2, \ldots, n$$

and the dissipation based on *Breuer–Onat* functional is computed from $(2.63)_1$ and $(2.62)_1$, respectively as:

$$\mathcal{L}_{12}(s_1, s_2) = -\frac{2}{\mathcal{K}_\infty} \sum_{i,j=1}^{n} \frac{\alpha_i^3 \alpha_j^3 \mu_i \mu_j}{B_i B_j} \mathrm{e}^{-(\alpha_i s_1 + \alpha_j s_2)} < 0$$

$$\mathcal{L}(s_1, s_2) = -2\ \mathcal{K}_\infty \sum_{i,j=1}^{n} \frac{A_i A_j}{\alpha_i \alpha_j} \mathrm{e}^{-(\alpha_i s_1 + \alpha_j s_2)} < 0$$

$$D_m(t) = \frac{2}{\mathcal{K}_\infty} \sum_{i,j=1}^{n} \frac{\alpha_i^3 \alpha_j^3 \mu_i \mu_j}{B_i B_j} \left\{ \int_0^\infty \int_0^\infty \mathrm{e}^{-(\alpha_i s_1 + \alpha_j s_2)}\ \mathbf{E}_r^t(s_1)\ \bullet \mathbf{E}_r^t(s_2)\ ds_1 ds_2 \right\}$$

which can be rewritten as a clearly nonnegative expression

$$\begin{aligned} D_m(t) &= \frac{2}{\mathcal{K}_\infty} \left(\sum_{i=1}^{n} \frac{\alpha_i^3 \mu_i}{B_i} \int_0^{+\infty} \mathrm{e}^{-\alpha_i s_1} \mathbf{E}_r^t(s_1)\ ds_1 \right) \\ &\quad \bullet \left(\sum_{l=1}^{n} \frac{\alpha_l^3 \mu_l}{B_l} \int_0^{+\infty} \mathrm{e}^{-\alpha_l s_2} \mathbf{E}_r^t(s_2)\ ds_2 \right) \\ &= \frac{2}{\mathcal{K}_\infty} \left(\sum_{i=1}^{n} \frac{\alpha_i^3 \mu_i}{B_i} \int_0^{+\infty} e^{-\alpha_i s} \mathbf{E}_r^t(s)\ ds \right)^2 \geq 0 \end{aligned}$$

2.5.4 *Implicit Constitutive Theories*

It could not be emphasized enough that the study of incompressible fluids whose viscosity is dependent on the pressure in addition to the well-known dependence on temperature is not a theoretical exercise, and there are many applications in engineering such as lubrication, film flows, and flows of granular materials where the dependence on pressure of the viscosity is crucial to model the flow. That the viscosity in the linear model of the response of fluids to stimuli he constructed could also depend on the pressure was recognized by Stokes himself back in 1845 [161]. In his remarkably foresighted paper Stokes discussed in fact in detail departures from the linearly constitutive model that bears his name. He goes on saying that *based on the experiments of Du Buat* [162], *which seem to show that increasing the pressure in channel and pipe flow of water does not increase the viscosity of the water he will assume that the viscosity of water and by analogy the viscosity of other incompressible fluids is independent of pressure*. We know now that the insight of Stokes was correct and there are situations when the variation of viscosity with pressure cannot be neglected and if ignored will lead to unacceptably large errors. In 1893, Barus [163] determined that at high pressures viscosity varies exponentially with pressure and proposed $\mu = \mu_0 \exp(\alpha\, p)$ for the relationship between viscosity and pressure. In this equation, α has units of Pa^{-1} and p is measured in Pa (Pascal). The dependence is indeed very substantial at lower pressures as well. α has been measured for various liquids an example of which is the Naphthalemic mineral oil, which shows a decreasing trend with increasing temperature, with temperature quadrupling from 20 to 80 °C α decreases from ~26 to ~16 GPa^{-1}, Höglund [164]. Subsequent attempts to improve on the Barus formula include those by Roelands [165], Irving and Barlow [166], and Paluch et al. [167]. Roelands' formula is given by:

$$\mu = \exp\left\{(\ln\omega_0 + 9.67)\left[-1 + \left(1 + 5.1 \times 10^{-9} p\right)^n\right]\right\}$$

where n is a constant is quite good away from the glass transition point. Irving and Barlow's empirical formula involves a double exponent and four temperature-dependent constants A, B, C, and D as:

$$\mu = \exp\{A\exp(Bp) - C\exp(-Dp)\}$$

Paluch's formula, which applies to certain low-molecular weight liquids, is similar to Barus' with the constant exponent α in Barus' formula replaced with $C\,(P_0 - p)^{-1}$ where C is an empirical constant.

In the first half of the last century, a tremendous amount of work was done on the response of fluids at high pressure and an exhaustive account of the literature up to 1931 can be found in the authoritative book by Bridgman [168] who had previously published extensive research on the variation of viscosity of great many fluids with pressure, Bridgman [169]. In particular, the work of Andrade [170] who suggested the following relationship for the viscosity η:

$$\eta(p, \rho, \theta) = A\sqrt{\rho}\, \exp\left[\left(p + r\rho^2\right) \frac{s}{\theta}\right]$$

where r, s, and A are constants and θ represents the temperature is noteworthy. However, as Andrade himself remarks it is not at all clear if such a model would work for *all* liquids in a certain temperature range although it would work for *a* class of fluids. A body of evidence that pressure-dependent viscosity is enormously important under certain physical situations kept accumulating in the second half of the last century as well, Griest et al. [171], Johnson and Cameron [172], Johnson and Tevaarwerk [173], Bair and Winer [174], and Bair and Kottke [175]. There are situations when a tenfold increase in pressure may lead to increases in viscosity of the order of $\boldsymbol{O}(10^8)$ %, whereas the density may change by a mere 10–20 % for the same tenfold increase in pressure inevitably leading to the conclusion that the change in density is insignificant, thus assuming that the fluid is incompressible, as compared to the change in viscosity that may happen for instance in elastohydrodynamics, Szeri [176]. The density variation with pressure p is well correlated by the following empirical equation, where ρ_0 is the density of the liquid as the pressure tends to zero, Dowson and Higginson [177] as:

$$\rho = \rho_0 \left[1 + \frac{0.6p}{1 + 1.4p} \right]$$

Clearly, the change in density ρ is negligible, of the order of 5 % when the pressure varies for example from 2 to 3 GPa.

Implicit relationships for the stress $\mathbf{T}$ of the form $\mathcal{F}(\mathbf{T}, \mathbf{D}, \theta) = 0$ defining the dependence of the stress on the deformation gradient $\mathbf{D}$ and temperature θ, of which Navier–Stokes equations are a special subclass, were studied by Rajagopal [178]. If one assumes that $\mathcal{F}$ is an isotropic tensor-valued function of the tensors $\mathbf{T}$ and $\mathbf{D}$, omitting the dependence on the temperature θ and the density ρ for simplification of the resulting expression, frame indifference and isotropy require that under the set of all orthogonal transformations O^+:

$$\mathcal{F}(\mathbf{QTQ}^{\mathrm{T}}, \mathbf{QDQ}^{\mathrm{T}}) = \mathbf{Q}\mathcal{F}(\mathbf{T}, \mathbf{D})\mathbf{Q}^{\mathrm{T}} \quad \forall \mathbf{Q} \in O^+$$

The theory of invariants then implies, Spencer [179]:

$$\begin{aligned} &\alpha_0 \mathbf{1} + \alpha_1 \mathbf{T} + \alpha_2 \mathbf{D} + \alpha_3 \mathbf{T}^2 + \alpha_4 \mathbf{D}^2 + \alpha_5 (\mathbf{DT} + \mathbf{TD}) + \alpha_6 (\mathbf{T}^2\mathbf{D} + \mathbf{DT}^2) \\ &\quad + \alpha_7 (\mathbf{TD}^2 + \mathbf{D}^2\mathbf{T}) + \alpha_8 (\mathbf{T}^2\mathbf{D}^2 + \mathbf{D}^2\mathbf{T}^2) = \mathbf{0} \end{aligned} \tag{2.68}$$

where all nine material functions α_i, $i = 0, \ldots, 8$ depend on the invariants $tr\mathbf{T}$, $tr\mathbf{D}$, $tr\mathbf{T}^2$, $tr\mathbf{D}^2$, $tr\mathbf{T}^3$, $\mathrm{tr}\mathbf{D}^3$, $tr\mathbf{TD}$, $tr\mathbf{T}^2\mathbf{D}$, $tr\mathbf{D}^2\mathbf{T}$, and $tr\mathbf{T}^2\mathbf{D}^2$. Rajagopal observes that density and temperature can be easily incorporated into this process by including them in the list of quantities on which the material functions depend. Now if one makes the following choices for the constants in (2.68):

$$\alpha_0 = \frac{1}{3} tr\mathbf{T}, \quad \alpha_1 = 1, \quad \alpha_2 = -\mu \, tr\mathbf{T}$$

a generalization of the Navier–Stokes equations with pressure-dependent viscosity is obtained.

$$\mathbf{T} = -p\mathbf{1} + 2\mu(p)\,\mathbf{D} \tag{2.69}$$

Existence of solutions to models of the type (2.69) has been proven for the space periodic case by Málek et al. [180] and the existence of weak solutions for steady flows under homogeneous Dirichlet boundary conditions and to specific body forces that are not necessarily small by Franta et al. [181]. The remarkable feature here in this implicit model formulation is that the constraint of incompressibility does not need to be enforced through Lagrange multipliers as is the case with the regular Navier–Stokes theory, or that the constraint stress does no work as it is routinely assumed in classical continuum mechanics. That this does not need to be the case in general was first recognized by Gauss [182] in a different context that of the motion of rigid bodies and was further explored recently in detail by Rajagopal [183] and Rajagopal and Srinivasa [184] who have shown that it is *unnecessary* within the context of continua to appeal to the assumption that the constraint stress is workless. Rajagopal observes that further generalizations of this approach, which would include *many of the rate-type models that are used to describe viscoelastic fluids*, can be achieved by selecting implicit relations of the form:

$$\mathcal{F}\left(\mathbf{T}, \dot{\mathbf{T}}, \ldots \overset{(n)}{\mathbf{T}}, \mathbf{D}, \dot{\mathbf{D}}, \ldots \overset{(n)}{\mathbf{D}}\right) = 0$$

where the superposed dot represents the material derivative and the superscript (n) stands for n material time derivatives; he further demonstrates how to derive further generalizations of the Navier–Stokes model, for instance such as

$$\mathbf{T} = -p\mathbf{1} + 2\left\{\mu\left(tr\mathbf{T}, |\mathbf{D}|^2\right)\right\}\mathbf{D}$$

among others. further extensions of the implicit approach would allow the construction of models for turbulent flows wherein the material functions can depend on the invariants associated with the stresses and their fluctuations as opposed to allowing them to only depend on the fluctuations in the velocity gradients.

Another way of approaching the problem would be for example to start with the assumption that the stress in the fluid depends on the density ρ, temperature θ, and the velocity gradient $\nabla\mathbf{u}$. Then it follows from frame-indifference and isotropy that the stress $\mathbf{T}$ in such a fluid is given by (2.70) often referred to as the Stokesian model:

$$\begin{aligned} \mathbf{T} &= \alpha_0(\rho, \theta, \mathrm{I}_\mathbf{D}, \mathrm{II}_\mathbf{D}, \mathrm{III}_\mathbf{D})\,\mathbf{1} + \alpha_2(\rho, \theta, \mathrm{I}_\mathbf{D}, \mathrm{II}_\mathbf{D}, \mathrm{III}_\mathbf{D})\,\mathbf{D} \\ &\quad + \alpha_4(\rho, \theta, \mathrm{I}_\mathbf{D}, \mathrm{II}_\mathbf{D}, \mathrm{III}_\mathbf{D})\,\mathbf{D}^2 \\ 2\mathbf{D} &= \nabla\mathbf{u} + (\nabla\mathbf{u})^\mathrm{T}, \quad \mathrm{I}_\mathbf{D} = tr\mathbf{D}, \quad 2\mathrm{II}_\mathbf{D} = (tr\mathbf{D})^2 - tr\mathbf{D}^2, \quad \mathrm{III}_\mathbf{D} = \det\mathbf{D} \end{aligned} \tag{2.70}$$

The term $\alpha_0(\rho,\theta,\mathrm{I}_\mathbf{D},\mathrm{II}_\mathbf{D},\mathrm{III}_\mathbf{D})$ does not have the meaning of mean normal stress as neither $tr\mathbf{D}$ nor $tr\mathbf{D}^2$ are zero. If one now assumes that the relationship between stress and strain is linear the Stokesian model reduces readily to by renaming the equation:

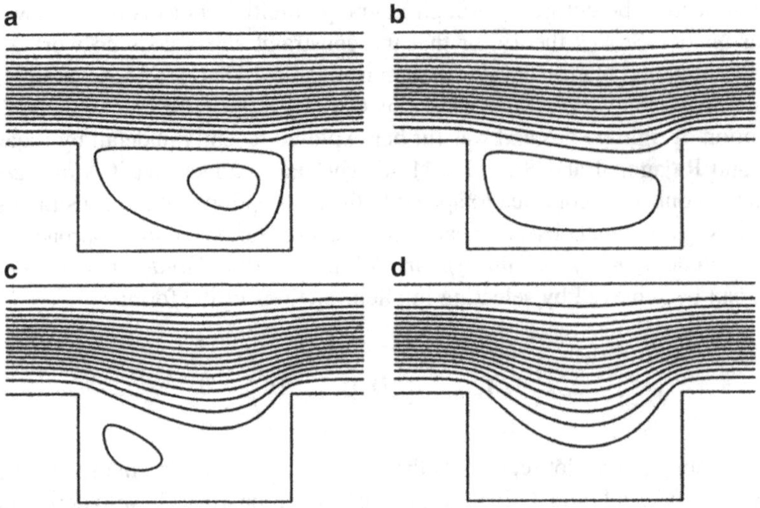

Fig. 2.9 Streamlines—flow over a slot—flow is from left to right: Constant viscosity (**a**) $\alpha = 0$; pressure dependent viscosity—Barus equation (**b**) $\alpha = 100$, (**c**) $\alpha = 200$, and (**d**) $\alpha = 360$ (Adapted from Hron et al. [183] with permission)

$$\mathbf{T} = \left[\widehat{\alpha}_o(\rho,\theta) + \lambda(\rho,\theta)\, tr\mathbf{D}\right]\mathbf{1} + 2\mu\,(\rho,\theta)\mathbf{D}$$

If the fluid is also incompressible, $tr\mathbf{D} = 0$, (2.70) reduces to the following:

$$\mathbf{T} = -p\mathbf{1} + \widehat{\alpha}_2(\theta, \mathrm{II}_\mathbf{D}, \mathrm{III}_\mathbf{D})\,\mathbf{D} + \widehat{\alpha}_4(\theta,\ \mathrm{II}_\mathbf{D}, \mathrm{III}_\mathbf{D})\,\mathbf{D}^2$$

as the stress $\mathbf{T}$ now depends only on the velocity gradient. Here, the pressure (p) is the Lagrange multiplier introduced to enforce incompressibility, $p = -1/3tr\mathbf{T}$.

Hron et al. [185] studied departures from Newtonian flow configurations in the same geometrical setting of incompressible fluids whose Cauchy stress is given by:

$$\mathbf{T} = -p\mathbf{1} + 2\mu\,(p)\,|\mathbf{D}|^{m-2}\mathbf{D} \tag{2.71}$$

which allows for shear-thinning $[m \in \langle -1, 2)]$ or shear-thickening $[m > 2]$. If $\mu(p)$ is a constant (2.71) represents the behavior of power law fluids. Hron et al. [185] study the behavior of these fluids when $\mu(p) = \alpha\, p^\gamma$, $\mu(p) = \exp(\alpha\, p)$ to show that that unidirectional Poiseuille flows are possible for linear dependence of the viscosity on the pressure $\mu(p) = \alpha\, p$, and explicit exact continuous solutions can be established in that case even if shear-thinning effects are included. For other forms of the viscosity, with polynomial and exponential dependence on the pressure, unidirectional flows are not possible. They also show that pressure gradient driven unidirectional flow between moving plates for fluids whose viscosity depends on certain fractional powers [$m = (n + 1)/n$, $n \in N$ where N is the set of positive integers] of the norm of the velocity gradient allow for multiplicity of solutions, a result not possible with the classical Navier–Stokes fluid and power-law fluids. Bair et al. [186] had shown that fluids of the type (2.71) with $m = 2$ are not capable of unidirectional flows and require a secondary flow. However, Hron et al. [185] show that this finding does not apply to all fluids with pressure-dependent viscosities as outlined above. They also study two steady plane flows numerically, flow between two coaxial, eccentric cylinders, and flow past a slot in a channel. In the former case, they find that although the pressure field differs from that for the Navier–Stokes solution by nearly two orders of magnitude, the velocity field is not markedly different. In the latter case, the dependence of the viscosity on the pressure can significantly change the structure of the flow field, Fig. 2.9.

Elastohydrodynamic lubrication calculations are routinely based on the classical Reynolds equation. Reynolds [187] derived the equation named after him under the assumption of constant viscosity, consistent with *Stokes' assumption that the viscosity is a constant for the Navier–Stokes* fluid. But to account for the increase in viscosity by several orders of magnitude due to the enormous pressures generated in an elastohydrodynamic contact, a pressure-dependent viscosity is inserted into the Reynolds equation a posteriori, thus leading to the neglect of some terms in the equation had the pressure-dependent viscosity been accounted for right from the outset in the derivation of the equation. This practice leads to a glaring inconsistency in the current elastohydrodynamic lubrication calculations, which went unnoticed until Rajagopal and Szeri [188] derived the elastohydrodynamic lubrication equation of Reynolds assuming at the very outset that the viscosity is pressure dependent. The Reynolds equation derived under the assumptions that the film is very thin and the body forces can be neglected reads as:

$$\frac{\partial}{\partial x}\left(\frac{h^3}{\mu}\frac{\partial p}{\partial x}\right) + \frac{\partial}{\partial z}\left(\frac{h^3}{\mu}\frac{\partial p}{\partial z}\right) = 6\widehat{U}\frac{\partial h}{\partial x} \tag{2.72}$$

where h is the film thickness and $\widehat{U}$ is the relative translation or rotation of the surfaces in the direction of the motion, the x direction. $\widehat{U} = U_{x(1)} - U_{x(2)}$ if surfaces are in relative translation like in the case of thrust bearings and $\widehat{U} = U_{x(1)} + U_{x(2)}$ if the surfaces are in relative rotation like in the case of journal bearings. The difficulties with (2.72) if the viscosity is pressure dependent are apparent right away because if that is the case the equations of motion cannot be integrated to obtain (2.72) in the first place. Rajagopal and Szeri [188] derived from the first principles the following set of equations with three unknowns, which governs the flow of the lubricating layer in the narrow gap.

$$\frac{\partial p}{\partial x}=\mu\frac{\partial^2 u}{\partial y^2}+\frac{d\mu}{dp}\left[\frac{\partial u}{\partial y}\frac{\partial p}{\partial y}+2\frac{\partial u}{\partial x}\frac{\partial p}{\partial x}\right],\quad \frac{\partial p}{\partial y}=\frac{d\mu}{dp}\frac{\partial u}{\partial y}\frac{\partial p}{\partial x},\quad \frac{\partial u}{\partial x}+\frac{\partial v}{\partial y}=0 \tag{2.73}$$

To proceed further requires the specification of the explicit dependence of viscosity on pressure. If an exponential dependence of the Barus type is assumed together with $\partial p/\partial y = 0$, integrating $(2.73)_1$, using boundary conditions $u\,|_{y=0} = U_{x(1)}$, $u\,|_{y=h(x)} = U_{x(2)}$ and substituting the result in $(2.73)_3$ leads to:

$$\frac{d}{dx}\left[\left(\frac{h^3}{\mu}-12\alpha\int_0^h y(h-y)\frac{\partial u}{\partial x}dy\right)\frac{dp}{dx}\right]=6\,\widehat{U}\frac{dh}{dx} \tag{2.74}$$

The modified lubrication equation (2.74) does differ from the classical Reynolds equation (2.72) because of the additional α term, which may lead to substantially different results.

Thermodynamical compatibility issues with implicit constitutive equation formulations were further explored by Rajagopal [183] who has shown that the model of an incompressible fluid with pressure-dependent viscosity is a natural consequence of requiring that the *Helmholtz potential* ψ depends only on the temperature θ while the rate of dissipation ξ depends on both the stress $\mathbf{T}$ and the symmetric part of the velocity gradient $\mathbf{D}$. In general, in an implicit theory, the Helmholtz potential can depend on the stress as well as on the kinematical quantities. Assuming a rate of dissipation of the form $\xi = 2[\mu\ (\theta,\ tr\mathbf{T})]\ \mathbf{D}\bullet\mathbf{D}$ and maximizing the rate of dissipation subject to $\xi = \mathbf{T}\bullet\mathbf{D} - \rho\,\dot{\psi} \geq 0$ as a constraint, where $\mathbf{T}\bullet\mathbf{D}$ is referred to as the stress power and $\dot{\psi} = \frac{d\psi}{dt}$, together with the constraint of incompressibility leads to a model of type (2.69) with μ constant. We recall that the usual procedure to derive (2.69) with constant μ is to split the stress into a constraint stress, which does no work, and a constitutively determined part. Assuming a more general form of the rate of dissipation $\xi = 2\ [\mu\ (\theta,\ tr\mathbf{T},\ \mathrm{II}_{\mathbf{D}})\ ]\ \mathbf{D}\bullet\mathbf{D} \geq 0$ and the requirement that the fluid is incompressible leads to models of the type $\mathbf{T} = -\ p\mathbf{1} + \mu\ (p,\theta,\mathrm{II}_{\mathbf{D}})\ \mathbf{D}$.

2.6 Maxwell-Like Differential Equations

A canonical form for a fairly general class of differential single mode Maxwell-like constitutive equations, which include the Johnson–Segalman [189], Gordon–Schowalter [190], Phan-Thien–Tanner [76], Phan-Thien [77], White–Metzner [191], upper-convected Maxwell, Giesekus [78, 79], Leonov [192], and Larson [193] as well as the molecular based FENE (Finite Extensible Non-linear Elastic) dumbbell model [194], can be constructed. Then all that is needed to obtain a particular model is to assign the relevant explicit form to the elastic potential. Leonov and co-workers show that the first three in this class, Johnson–Segalman, Gordon–Schowalter, and the non-affine versions of the Phan-Thien–Tanner model:

$$f[tr\mathbf{S}]\ \ \mathbf{S} + \lambda \overset{\circ}{\mathbf{S}} = 2\eta\mathbf{D}, \quad \mathbf{T} = -p\mathbf{1} + \mathbf{S},$$

$$\overset{\circ}{\mathbf{S}} = \overset{\bullet}{\mathbf{S}} - \left(\nabla\mathbf{u}^{\mathrm{T}} - \xi\mathbf{D}\right)\mathbf{S} - \mathbf{S}\left(\nabla\mathbf{u}^{\mathrm{T}} - \xi\mathbf{D}\right)^{\mathrm{T}}$$

$$\overset{\circ}{\mathbf{S}} = \overset{\bullet}{\mathbf{S}} - \zeta^{\mathrm{T}}\mathbf{S} - \mathbf{S}\zeta - \alpha(\mathbf{DS} + \mathbf{SD}), \quad \alpha = 1 - \xi$$

where $\overset{\circ}{\mathbf{S}}$ if the Gordon–Schowalter mixed convected derivative ($\alpha \neq 1, 0, -1$), are Hadamard unstable. However, the original affine version of the upper-convected Phan-Thien–Tanner equation ($\alpha = 1 : \overset{\circ}{\mathbf{S}} \to \overset{\nabla}{\mathbf{S}}$) is both Hadamard and dissipative stable. White–Metzner model [191]:

$$S_{kl} + \lambda\ \ (\mathrm{II}_{\mathbf{D}})\overset{\nabla}{S}_{kl} = 2\eta\ \ (\mathrm{II}_{\mathbf{D}})D_{kl}, \quad \mathrm{II}_{\mathbf{D}} = 2\mathbf{D} : \mathbf{D}$$

$$\lambda\,(\mathrm{II}_{\mathbf{D}}) = \frac{\lambda_0}{1 + a\lambda_0(2\mathbf{D} : \mathbf{D})^{1/2}}$$

is both Hadamard and dissipative unstable, and the upper-convected Maxwell model:

$$S_{kl} + \lambda \overset{\nabla}{S}_{kl} = 2\mu\ D_{kl}$$

$$\overset{\nabla}{\mathbf{S}} = \frac{D\mathbf{S}}{Dt} - \nabla\mathbf{u}^{\mathrm{T}}\mathbf{S} - \mathbf{S}\nabla\mathbf{u} = \overset{\bullet}{\mathbf{S}} - \mathbf{DS} - \mathbf{SD}$$

is globally Hadamard stable but dissipative unstable. In the above $\mathbf{S}$, $\mathbf{D}$, $\mathbf{T}$, $\mathbf{1}$, ζ, $\mathbf{u}$, λ, η, and ξ represent the extra-stress tensor, the rate of deformation tensor, the total stress tensor, the unit tensor, the vorticity tensor, the velocity vector, the relaxation time, the shear viscosity, and the slippage factor, respectively. The second invariant of the rate of deformation tensor $\mathbf{D}$ is represented by $\mathrm{II}_{\mathbf{D}}$, and $(\overset{\bullet}{\bullet})$ and $(\overset{\nabla}{\bullet})$ refer to the material derivative and the upper-convected derivative, respectively. Giesekus [78, 79]:

$$\mathbf{S} + \lambda \overset{\nabla}{\mathbf{S}} + \alpha\frac{\lambda}{\eta_p}\mathbf{S}^2 = \eta_p\mathbf{D}$$

where η_{p} is the polymer contributed viscosity and the original Leonov equation [192] are dissipative unstable, Larson [193] is Hadamard unstable, and the FENE equation [194] is globally Hadamard and dissipative stable. All in all, there are three Maxwell-like differential constitutive equation specifications which are stable both in the sense of Hadamard and dissipative stability: FENE, the original affine upper-convected Phan-Thien–Tanner and later versions of the Leonov equation. However, it is worth noting that the first two predict zero second normal stress difference in simple shear flow and as a consequence cannot be used to predict secondary flows (see Siginer [29], Sect. 3.4).

2.7 Single Integral Constitutive Equations

Rivlin and Sawyers [195] presented the canonical form of the single integral type constitutive equations for incompressible viscoelastic fluids. The extra-stress tensor **S** and the evolution equation for the Cauchy strain tensor **C** in canonical form are expressed as:

$$\mathbf{S} = \int_{-\infty}^{t} \left\{ \varphi_1\left(\mathrm{I_C}, \mathrm{I_{C^{-1}}}, t-\tau\right) \mathbf{C}_t(\tau) - \varphi_2\left(\mathrm{I_C}, \mathrm{I_{C^{-1}}}, t-\tau\right)\ \mathbf{C}_t^{-1}(\tau) \right\}\, d\tau, \tag{2.75}$$

$$\mathrm{I_C} = tr\mathbf{C}, \quad \mathrm{I_{C^{-1}}} = tr\mathbf{C}^{-1}$$

$$\overset{\nabla}{\mathbf{C}} = \frac{\partial \mathbf{C}}{\partial t} + \mathbf{u} \bullet \nabla \mathbf{C} \;-\; \mathbf{C} \bullet \nabla \mathbf{u} - \;(\nabla \mathbf{u})^{\mathrm{T}} \bullet \mathbf{C} = \mathbf{0}, \quad \mathbf{C}|_{t=\tau} = \mathbf{1}$$

where $\mathbf{C}_t^{-1}(\tau)$, **1**, t, and τ represent the relative Finger deformation tensor, the unit tensor, the present time, and the past time, respectively. Based on experimental evidence, time–strain separable kernels can be introduced as:

$$\varphi_n\left(\mathrm{I_C}, \mathrm{I_{C^{-1}}}, t-\tau\right) = \frac{dG(t-\tau)}{d\tau}\ \hat{\varphi}_n\left(\mathrm{I_C}, \mathrm{I_{C^{-1}}}\right), \qquad n = 1, 2$$

with $G(t)$ representing the relaxation modulus. Assigning explicit forms to the kernels φ_n in the non-separable case and to $\hat{\varphi}_n$ in the separable case leads to different types of equations proposed in the literature that found favor with the practitioner. For instance, the well-known Lodge model, the integral representation of the upper-convected Maxwell equation, is obtained when $\hat{\varphi}_1 = 1, \quad \hat{\varphi}_2 = 0$. The Lodge equation is dissipative unstable. The separable and the non-separable K-BKZ models of which Lodge equation is a special case are arrived at when the kernels are expressed in terms of the elastic Hookean modulus $G(\theta)$ a function of the temperature θ and of the relaxation effects dependent and independent of the thermodynamic free energy potential (Helmholtz potential) $\hat{\psi}$ and $\psi(t)$ in the separable and non-separable time strain–rate cases, respectively:

$$\hat{\varphi}_1 = \frac{2\rho}{G}\frac{\partial \hat{\psi}}{\partial \mathrm{I_C}}, \quad \hat{\varphi}_2 = \frac{2\rho}{G}\frac{\partial \hat{\psi}}{\partial \mathrm{I_{C^{-1}}}}, \quad \varphi_1 = \frac{2\rho}{G}\frac{\partial \psi(t)}{\partial \mathrm{I_C}}, \quad \varphi_2 = \frac{2\rho}{G}\frac{\partial \psi(t)}{\partial \mathrm{I_{C^{-1}}}}$$

The separable K-BKZ class is dissipative unstable. The Larson and Monroe model [196], which falls in this category:

$$\hat{\varphi}_1 = \frac{2\rho}{G}\frac{\partial\hat{\psi}}{\partial I_C} = \frac{\partial}{\partial I_C}\left\{\frac{3}{2F_1(I_C, I_{C^{-1}})}\ln\left\{1 + \frac{F_1(I_C, I_{C^{-1}})\,[F_2(I_C, I_{C^{-1}}) - 3]}{3}\right\}\right\}$$

$$\hat{\varphi}_2 = \frac{2\rho}{G}\frac{\partial\hat{\psi}}{\partial I_{C^{-1}}} = \frac{\partial}{\partial I_{C^{-1}}}\left\{\frac{3}{2F_1(I_C, I_{C^{-1}})}\ln\left\{1 + \frac{F_1(I_C, I_{C^{-1}})\,[F_2(I_C, I_{C^{-1}}) - 3]}{3}\right\}\right\}$$

$$F_1 = a + b\ \tanh^{-1}\left\{\frac{cF_3^3}{1 + F_3^2}\right\}$$

$$F_2 = (1 - \beta)I_C + \sqrt{1 + 2\beta I_{C^{-1}}} - 1, \quad F_3 = I_{C^{-1}} - I_C$$

is also Hadamard unstable in addition to being dissipative unstable. Here, the parameters a, b, and c are curve fitting parameters. The popular Wagner equations, the original (Wagner I) [197] and the modified factorable model (Wagner II) [198] proposed later, come out of (2.75) when the kernels are defined respectively as:

$$\hat{\varphi}_1 = a\ \exp(-bf) + (1 - a)\ \exp(-cf), \quad f = \beta I_C + (1 - \beta)I_{C^{-1}}, \quad \hat{\varphi}_2 = 0,$$

$$\hat{\varphi}_1 = (1 - \beta)f_1(I_C, I_{C^{-1}}), \quad \hat{\varphi}_2 = \beta f_1(I_C, I_{C^{-1}}),$$

$$f_1(I_C, I_{C^{-1}}) = \frac{1}{\alpha\sqrt{(I_C - 3)(I_{C^{-1}} - 3)}}$$

In both of these kernel specifications, a, b, c, α, and β are positive curve fitting parameters. Both Wagner models are Hadamard unstable. Papanastasiou et al. [199] proposed:

$$\hat{\varphi}_1 = \beta f_2(I_C, I_{C^{-1}}), \quad \hat{\varphi}_2 = 0, \quad f_2(I_C, I_{C^{-1}}) = [\beta - 3 + \alpha I_C + (1 - \alpha)I_{C^{-1}}]^{-1}$$

to which Luo and Tanner [200] suggested the following modification:

$$\hat{\varphi}_2 = \chi\beta f_2(I_C, I_{C^{-1}}), \quad \chi > 0$$

Again β, α, and χ are numerical curve fitting parameters with experimental data. Both Papanastasiou and Luo and Tanner models are Hadamard unstable. It is surprising to note that none of the single integral time–strain separable kernel type of constitutive equations is evolutionary, that is, none is both Hadamard and dissipative stable, and the predictions of some, the derivation of which is partly based on a molecular approach such as the Larson model, is worse than others. We note that in any variant of the Rivlin–Sawyers family of CEs if $\varphi_2 = 0$ or in separable cases if $\hat{\varphi}_2 = 0$ the second normal stress difference will be zero and therefore no secondary flows can be predicted in non-circular tube flows (see Siginer [29], Sect. 3.4) even though the CE may be Hadamard and dissipative stable.

2.8 Hadamard Instability

Hadamard instability is well understood in non-linear elasticity. Stability criteria in the Hadamard sense are equivalent to the conditions of strong ellipticity of the field equations for dynamic problems in non-linear elasticity. If the solution of the field equations for a specific Cauchy boundary value problem provides the initial conditions for marching the solution at subsequent times in the time line, the set of field equations are called Hadamard stable or evolutionary or well posed. In addition, there are thermodynamic stability criteria called the GCN^{+} conditions, which require the convexity of the thermodynamic elastic potential with respect to the strain measure (Hencky strain), Truesdell and Noll [201]. It turns out that the thermodynamic stability conditions can be interpreted as *necessary* conditions for Hadamard stability. But Leonov [26] shows that the conditions he derives for Hadamard stability for the class of Maxwell-like differential equations and time–strain separable single integral constitutive equations, which include almost all the equations in popular use, are inclusive of those for GCN^{+} stability. That is they are stronger. Although Hadamard instability for viscoelastic CEs was first studied by Rutkevich [202, 203] as far back as 1969, the publications were in Russian and remained relatively unknown in the West until recently. The equivalent of the conditions for Hadamard stability in viscoelasticity was rigorously worked out by Leonov and his co-workers in a sustained effort in the early 1990s, Leonov [204] and Kwon and Leonov [205, 206]. They coincide roughly with the conclusions drawn in the context of non-linear elasticity. Namely when the extra-stress **S** is expressed in terms of an elastic potential $W(tr\mathbf{C}, tr\mathbf{C}^{-1}, \theta) = \rho\psi$, the strain energy function for incompressible materials with θ, ρ, ψ, and **C** representing the temperature, density, Helmholtz free energy, and the Finger tensor, respectively, elastic potential W must satisfy the convexity conditions:

$$\frac{\partial W}{\partial I_C} > 0, \quad \frac{\partial W}{\partial I_{C^{-1}}} > 0, \quad \frac{\partial^2 W}{\partial I_C^2}\frac{\partial^2 W}{\partial I_{C^{-1}}^2} > \frac{\partial^2 W}{\partial I_C \partial I_{C^{-1}}} \tag{2.76}$$

which are sufficient for Hadamard stability of the set of constitutive and evolution equations defining **S**. These conditions are equivalent to stating that the thermodynamic potential F is a monotonically increasing convex function of the invariants $tr\mathbf{C}$ and $tr\mathbf{C}^{-1}$.

The consequences of Hadamard instability are devastating. As the solution cannot be continued or as well is not continuous along the time line very quick blow-up instabilities with very short-wave disturbances occur with devastating results for numerical computations which fail to converge. Refining the computational mesh to improve the results actually worsens them because of the extremely short wave nature of the disturbances. Hadamard instability can be interpreted in many cases as due to the amplitude of the initially infinitesimal waves tending to infinity as the wavelength tends to zero (see also Sect. 1.2). If the constitutive equation has the ability to provide a non-linear rapid response, the instability may be smoothed over. This association with the non-linear rapid response of the constitutive equation makes the Hadamard instability dependent on the properties of the type of differential operator in the evolution equation for differential models, Leonov [26]. One can conclude that for stability, the initially small (infinitesimal) amplitude of disturbing waves should remain small at all times. An excellent example of Hadamard stability analysis for the Oldroyd-B and UCM fluids can be found in Owens and Phillip [81], p. 55–59.

2.9 Dissipative Instability

Every constitutive equation includes a dissipative term or terms. The formulation of the non-equilibrium terms may give rise to another type of instability called dissipative instability. This is a relatively new issue in the study of instabilities which plague non-linear viscoelastic constitutive equations and was initiated by Leonov [204]. A constitutive equation may be both Hadamard and dissipative unstable, and a constitutive equation which is Hadamard stable may turn out to be dissipative unstable. Examples are the White–Metzner and the upper-convected Maxwell models. The latter is Hadamard stable in a global sense, but is dissipative unstable because it shows unbounded growth of stress in simple extension when the elongation rate exceeds the half of the reciprocal relaxation time. The former is Hadamard unstable (non-evolutionary) because of the dependence of the relaxation time on the invariant of the deformation rate tensor, Dupret and Marchal [207], the same dependence which gives the model its capability to predict shear-thinning behavior. The model is also dissipative unstable for a reason similar to that of the upper-convected Maxwell model. Specifically whenever the extensional strain rate exceeds the half of the reciprocal relaxation time, a dissipative instability occurs, Verdier and Joseph [208]. Necessary and sufficient conditions for dissipative stability for both the class of Maxwell-like and the class of single integral equations with time–strain separable kernels were determined in a series of papers by Leonov and co-workers, Leonov [204] and Kwon and Leonov [205, 206, 209].

Dissipative stability requires that firstly in any flow the free energy and dissipation functionals remain bounded and secondly steady flow curves in simple shear and in simple elongation must be monotonically and unboundedly increasing with respect to the strain rate. The former condition is framed explicitly with the following statement: The free energy and dissipation functionals will remain bounded if and only if the Hadamard stable elastic potential function $W(H_k)$ expressed in terms of the principal values H_k of the Hencky strain measure **H** does not grow exponentially:

$$\mathrm{H} = \frac{1}{2}\ \ln \mathrm{C}, \quad \mathrm{I}_{\mathbf{H}} = 0$$

The study of dissipative instabilities is far from being a mature subject and is very much work in progress as well as the whole field of instabilities related to viscoelastic constitutive equations even though considerable progress has been made in the last two decades and some light has been shed on the perplexing problem of blow-up instabilities in numerical computations at high Deborah or high Weissenberg numbers. It is also fair to conclude based on available studies that a CE which is dissipative stable is also very likely to be Hadamard stable, but the converse is not necessarily true. It is well known that it is possible to stabilize a Hadamard unstable constitutive equation by adding a Newtonian term. However, it is less well known that sometimes a Hadamard stable but dissipative unstable CE can be stabilized by changing the elastic potential. For instance, the single mode evolution equation for the elastic extra-stress **S** suggested by Giesekus in its widely used form:

$$\mathbf{S}\left[\mathbf{1}+\frac{\alpha\lambda}{\eta}\mathbf{S}\right]+\lambda\overset{\nabla}{\mathbf{S}}=2\eta\mathbf{D}$$

is dissipative unstable, but can be stabilized in the dissipative sense if an appropriate elastic potential is chosen to satisfy the convexity conditions (2.76). Here, $\alpha < 0.5, \lambda$ and η are constitutive parameters representing the ratio of the second normal stress difference to the first normal stress difference, the relaxation time and the viscosity.

Leonov [26] proposed on rigorous grounds a *robust Hadamard and dissipative stable* constitutive equation, which works well with melts and concentrated solutions at high *De* numbers of the order of several hundreds. The extra-stress is expressed in terms of an elastic potential $W(tr\mathbf{C}, tr\mathbf{C}^{-1}, \theta) = \rho\psi$, the strain energy function for incompressible materials is given by:

$$\mathbf{S}=2\left[\frac{\partial W}{\partial \mathrm{I_C}}\mathbf{C}-\frac{\partial W}{\partial \mathrm{I_{C^{-1}}}}\mathbf{C}^{-1}\right]$$

with θ, ρ, ψ, and **C** representing the temperature, density, Helmholtz free energy, and the Finger tensor, respectively. For Hadamard stability, elastic potential W must satisfy the convexity conditions (2.76). The evolution equation for the Cauchy strain measure **C** is given by:

$$2\lambda\overset{\nabla}{\mathbf{C}}+\;a\;\left\{\mathrm{I_C},\mathrm{I_{C^{-1}}}\right\}\left\{\mathbf{C}^2+\frac{(\mathrm{I_{C^{-1}}}-\mathrm{I_C})}{3}\;\mathbf{C}-\mathbf{1}\right\}=0$$

The dissipative function a is positive definite and has a linear viscoelastic limit. A general elastic potential is proposed:

$$W(\mathrm{I_C},\mathrm{I_{C^{-1}}},\theta)=\frac{3G(\theta)}{2(n+1)}(1-\beta)\left\{\left(\frac{tr\mathbf{C}}{3}\right)^{n+1}+\left(\frac{tr\mathbf{C}^{-1}}{3}\right)^{n+1}-2\right\}$$

which reduces the explicit form for the extra-stress **S** to:

$$\mathbf{S} = (1-\beta)\left(\frac{tr\mathbf{C}}{3}\right)^n \mathbf{C} - \beta\left(\frac{tr\mathbf{C}^{-1}}{3}\right)^n \mathbf{C}^{-1}$$

$G(\theta)$ is the linear Hookean elastic modulus and β and n are numerical fitting parameters. These equations *are Hadamard and dissipative stable* for $0 \leq \beta \leq 1$ and $n > 0$, and according to Leonov and co-workers have the ability to predict the behavior of concentrated solutions and melts at very high *De* numbers of the order of several hundreds and perhaps $De \sim 1{,}000$. The use of these equations in predicting secondary flows will be explored in Siginer [29].

References

1. Ruggeri T. Can constitutive relations be represented by non-local equations? Q Appl Math. 2012;70(3):597–611.
2. Eringen AC. Non-local continuum field theories. New York: Springer; 2002.
3. Fick A. Über diffusion. Ann Phys. 1855;94:59–86.
4. Darcy H. Les Fontaines Publiques de La Ville de Dijon. Paris: Victor Dalmont; 1856.
5. Rajagopal KK. On a hierarchy of approximate models for flows of incompressible fluids through porous solids. Math Model Meth Appl Sci. 2007;17(2):215–52.
6. Coleman B, Noll W. The thermomechanics of elastic materials with heat conduction and viscosity. Arch Ration Mech Anal. 1963;13:167–78.
7. Müller I. On the entropy inequality. Arch Ration Mech Anal. 1967;26:118–41.
8. Müller I. On the frame dependence of stress and heat flux. Arch Ration Mech Anal. 1972;45:241–50.
9. Truesdell C. Correction of two errors in the kinetic theory of gases which have been used to cast unfounded doubt upon the principle of material frame-indifference. Meccanica. 1976;11(4):196–9.
10. Bressan A. On relativistic heat conduction in the stationary and non-stationary cases, the objectivity principle and piezoelasticity. Lett Nuovo Cimento. 1982;33(4):108–12.
11. Ruggeri T. Generators of hyperbolic heat equation in non-linear thermoelasticity. Rend Sem Mat Univ Padova. 1982;68:79–91.
12. Eringen AC, Okada K. A lubrication theory for fluids with microstructure. Int J Eng Sci. 1995;33(15):2297–308.
13. Israelashvili JN. Measurement of the viscosity of liquids in very thin films. J Colloid Interface Sci. 1986;110:263.
14. Israelashvili JN. Measurement of the viscosity of thin fluid films between two surfaces with and without adsorbed polymers. Colloid Polymer Sci. 1986;264:1060–5.
15. Eringen AC. On non-local fluid mechanics. Int J Eng Sci. 1972;10(6):561–75.
16. Eringen AC. On non-local micro-fluid mechanics. Int J Eng Sci. 1973;11(2):291–306.
17. Speziale C, Eringen AC. Non-local fluid mechanics description of wall turbulence. Comput Math Appl. 1981;7:27–42.
18. Demiray H, Eringen AC. On non-local diffusion of gases. Arch Mech. 1978;30:65–77.
19. Eringen AC. Non-local inviscid magneto-hydrodynamics and dispersion of Alfven waves. Bull Tech Univ Istanbul. 1986;39:393–408.
20. McCay BM, Narasimhan MNL. Theory of non-local electromagnetic fluids. Arch Mech. 1981;33(3):365–84.
21. Narasimhan MNL, McCay BM. Dispersion of surface waves in non-local dielectric fluids. Arch Mech. 1981;33(3):385–400.

22. Speziale CG. On turbulent secondary flows in pipes of non-circular cross-section. Int J Eng Sci. 1982;20(7):863–72.
23. Speziale CG. PhD Thesis, Princeton University; 1978.
24. Reddy JN. Non-local theories for bending, buckling and vibration of beams. Int J Eng Sci. 2007;45:288–307.
25. Thai H-T. A non-local beam theory for bending, buckling, and vibration of nanobeams. Int J Eng Sci. 2012;52:56–64.
26. Leonov AI. Constitutive equations for viscoelastic liquids: formulation, analysis and comparison with data. In: Siginer DA, De Kee D, Chhabra RP, editors. Advances in the flow and Rheology of non-Newtonian fluids, part a, Rheology series 8. New York: Elsevier; 1999. p. 577–91.
27. Oldroyd JG. On the formulation of rheological equations of state. Proc Roy Soc Lond A. 1950;200:523–91.
28. Oldroyd JG. An approach to non-Newtonian fluid mechanics. J Non-Newton Fluid Mech. 1984;14:9–46.
29. Siginer DA. Dynamics of tube flow of viscoelastic fluids. New York: Springer; 2014.
30. Maxwell JC. On the dynamical theory of gases. Phil Trans Roy Soc Lond A. 1866;157:26–78.
31. Boltzmann L. Zür theorie der elästichen nachwirkung. Sitzgsberg Akad Wiss Wien. 1874;70:275–306.
32. Jeffreys H. The earth. Cambridge: Cambridge University Press; 1929.
33. Leonov AI. On a class of constitutive equations for viscoelastic liquids. J Non-Newton Fluid Mech. 1987;25(1):1–59.
34. Costa Mattos HS. A thermodynamically consistent constitutive theory for fluids. Int J Non Lin Mech. 1998;33(1):97–110.
35. Rajagopal KR, Srinivasa AR. A thermodynamic framework for rate-type fluid models. J Non-Newton Fluid Mech. 2000;88(3):207–27.
36. Fröhlich H, Sack R. Theory of the rheological properties of dispersions. Proc Roy Soc London A. 1946;185:415–30.
37. Oldroyd JG. The elastic and viscous properties of emulsions and suspensions. Proc Roy Soc London A. 1953;218:122–32.
38. Barnes HA, Hutton JF, Walters K. An introduction to Rheology. New York: Elsevier; 1989.
39. Burgers JM. Mechanical considerations-model systems-phenomenological theories of relaxation and of viscosity, first report on viscosity and plasticity, Nordermann publishing company, New York, 1935. 2nd edition prepared by the committee for the study of viscosity of the Academy of Sciences at Amsterdam. New York: Nordermann Publishing Company; 1939.
40. Burgers JM. Non-linear relations between viscous stresses and instantaneous rate of deformation as a consequence of slow relaxation. Proc Kon Ned Akad Wet. 1948;51:787–92.
41. Murali Krishnan J, Rajagopal KR. Review of the uses and modeling of bitumen from ancient to modern times. Appl Mech Rev. 2003;56:149–214.
42. Murali Krishnan J, Rajagopal KR. On the mechanical behavior of asphalt. Mech Mater. 2005;37(11):1085–100.
43. Quintanilla R, Rajagopal KR. On Burgers fluids. Math Meth Appl Sci. 2006;29:2133–47.
44. Quintanilla R, Rajagopal KR. Further mathematical results concerning Burgers fluids and their generalizations. Z Angew Math Phys. 2012;63:191–202.
45. Green MS, Tobolsky AV. A new approach to the theory of relaxing polymeric media. J Chem Phys. 1946;14(2):80–92.
46. Lodge AS. A network theory of flow birefringence and stress in concentrated polymer solutions. Trans Faraday Soc. 1956;52:120–30.
47. Lodge AS. Elastic liquids. New York: Academic; 1964.
48. Yamamoto M. The viscoelastic properties of network structure I. General formalism. J Phys Soc Japan. 1956;11:413–21.

49. Yamamoto M. The viscoelastic properties of network structure II. Structural viscosity. J Phys Soc Japan. 1957;12:1148–58.
50. Yamamoto M. The viscoelastic properties of network structure III. Normal stress effect (Weissenberg effect). J Phys Soc Japan. 1958;13:1200–11.
51. De Gennes PG. Scaling concepts in polymer physics. Ithaca, NY: Cornell University; 1979.
52. Doi M, Edwards SF. The theory of polymer dynamics. Oxford: Clarendon; 1986.
53. Kaye A. College of Aeronautics, Cranfield, Note No.134; 1962.
54. Bernstein B, Kearsley AE, Zapas L. A study of stress relaxation with finite strain. Trans Soc Rheol. 1963;7(1):391–410.
55. Marrucci G, Grizzuti N. The Doi-Edwards model in slow flows. Predictions on the Weissenberg effect. J Non-Newton Fluid Mech. 1986;21(3):319–28.
56. Marrucci G. The Doi-Edwards model without independent alignment. J Non-Newton Fluid Mech. 1986;21(3):329–36.
57. Wagner MH. A constitutive analysis of uniaxial elongational flow data of a low-density polyethylene melt. J Non-Newton Fluid Mech. 1978;4(1–2):39–55.
58. Wagner MH. Elongational behavior of polymer melts in constant elongation-rate, constant tensile stress, and constant tensile force experiments. Rheol Acta. 1979;18(6):681–92.
59. Tanner RI. Engineering Rheology (revised edition). Oxford: Clarendon; 1988.
60. Schowalter WR. Mechanics of non-Newtonian fluids. London: Pergamon Press; 1978.
61. Larson RG. Constitutive equations for polymer melts and solutions. Boston: Butterworth; 1988.
62. Bird RB, Curtiss CF, Armstrong RC, Hassager O. Dynamics of polymeric liquids, vol. 1. 2nd ed. New York: Wiley; 1987.
63. Carreau PJ, De Kee DCR, Chhabra RP. Rheology of polymeric systems. Munich, Vienna, NY: Hanser Publishers; 1997.
64. Bird RB, Dotson PJ, Johnson NL. Polymer solution rheology based on a finitely extensible bead-spring chain model. J Non-Newton Fluid Mech. 1980;7:213–35.
65. Chilcott MD, Rallison JM. Creeping flow of dilute polymer solutions past cylinders and spheres. J Non-Newton Fluid Mech. 1988;29:381–432.
66. Ghosh I, McKinley GH, Brown RA, Armstrong RC. Deficiencies of FENE dumbbell models in describing the rapid stretching of dilute polymer solutions. J Rheol. 2001;45:721–58.
67. Chabbra RP, Uhlherr PHT, Boger DV. The influence of fluid elasticity on the drag coefficient for creeping flow around a sphere. J Non-Newton Fluid Mech. 1980;6:187–99.
68. Sizaire R, Legat V. Finite element simulation of a filament stretching extensional rheometer. J Non-Newton Fluid Mech. 1997;71:89–107.
69. Bird RB, Wiest JM. Constitutive equations for polymeric liquids. Annu Rev Fluid Mech. 1995;27:169–93.
70. Warner HR. Kinetic theory and rheology of dilute suspensions of finitely extensible dumbbells. Ind Eng Chem Fund. 1972;11:379–87.
71. Peterlin A. Hydrodynamics of macromolecules in a velocity field with longitudinal gradient. J Polym Sci B. 1966;4:287–91.
72. Chandrasekhar S. Stochastic problems in physics and astronomy. Rev Mod Phys. 1943;15:1–89.
73. Lielens G, Halin P, Jaumain I, Keunings R, Legat V. New closure approximations for the kinetic theory of finitely extensible dumbbells. J Non-Newton Fluid Mech. 1998;76:249–79.
74. Lielens I, Keunings R, Legat V. The FENE-L and FENE-LS closure approximations to the kinetic theory of finitely extensible dumbbells. J Non-Newton Fluid Mech. 1999;87:179–96.
75. Verleye V, Dupret F. Prediction of fiber orientation in complex molded parts. In: Altan CAE, Siginer DA, Van Arsdale WE, Alexandrou AN, editors. Developments in non-Newtonian flows. New York: ASME; 1993. p. 139–63.
76. Phan-Thien N, Tanner RI. New constitutive equation derived from network theory. J Non-Newton Fluid Mech. 1977;2:353–65.
77. Phan-Thien N. Non-linear network viscoelastic model. J Rheol. 1978;22:259–83.

78. Giesekus H. A simple constitutive equation for polymer fluids based on the concept of deformation-dependent tensorial mobility. J Non-Newton Fluid Mech. 1982;11:69–109.
79. Giesekus H. A unified approach to a variety of constitutive models for polymer fluids based on the concept of configuration-dependent molecular mobility. Rheol Acta. 1982;21 (4–5):366–75.
80. Peters GWM, Schoonen JFM, Baaijens FPT, Meijer HEH. On the performance of enhanced constitutive models for polymer melts in a cross-slot flow. J Non-Newton Fluid Mech. 1999;82(2–3):387–427.
81. Owens RG, Phillips PN. Computational rheology. London: Imperial College Press; 2002.
82. Truesdell C, Noll W. The non-linear field theories of mechanics. 2nd ed. Berlin: Springer; 1992.
83. Reiner M. A mathematical theory of dilatancy. Am J Math. 1945;67:350–62.
84. Reiner M. Elasticity beyond the elastic limit. Am J Math. 1948;70:433–46.
85. Reiner M. Relations between stress and strain in complicated systems. Proc Int Con Rheol. 1949;1:1–21.
86. Rivlin RS. Large elastic deformations of isotropic materials, I. Fundamental concepts. Phil Trans Roy Soc Lond A. 1948;240:459–90.
87. Rivlin RS. Large elastic deformations of isotropic materials, IV., further developments of the general theory. Phil Trans Roy Soc Lond A. 1948;241:379–97.
88. Rivlin RS. Large elastic deformations of isotropic materials, V., the problem of flexure. Proc Roy Soc Lond A. 1949;195:463–73.
89. Rivlin RS. Large elastic deformations of isotropic materials, VI., further results in the theory of torsion, shear and flexure. Phil Trans Roy Soc Lond A. 1949;242:173–95.
90. Rivlin RS. The hydrodynamics of non-Newtonian fluids, I. Proc Roy Soc Lond A. 1948;193:260–81.
91. Rivlin RS. The hydrodynamics of non-Newtonian fluids, II. Proc Camb Phil Soc. 1949;45:88–91.
92. Rivlin RS. The normal stress coefficient in solutions of macro-molecules. Trans Faraday Soc. 1949;45:739–48.
93. Criminale Jr WO, Ericksen JL, Filbey GL. Steady shear flow of non-Newtonian fluids. Arch Ration Mech Anal. 1957;1(1):410–7.
94. Wang CC. A representation theorem for the constitutive equation of a simple material in motions with constant stretch history. Arch Ration Mech Anal. 1965;20:329–40.
95. Larson RG. Flows of constant stretch history for polymeric materials with power-law distributions of relaxation times. Rheol Acta. 1985;24:443–9.
96. Tanner RI, Huilgol RR. On a classification scheme for flow fields. Rheol Acta. 1975;14:959–62.
97. Green AE, Rivlin RS. The mechanics of non-linear materials with memory, part 1. Arch Ration Mech Anal. 1957;1:1–21.
98. Green AE, Rivlin RS, Spencer AJM. The mechanics of non-linear materials with memory, part 2. Arch Ration Mech Anal. 1959;3:82–90.
99. Green AE, Rivlin RS. The mechanics of non-linear materials with memory, part 3. Arch Ration Mech Anal. 1960;4:387–404.
100. Coleman BD, Noll W. An approximation theorem for functionals with applications in continuum mechanics. Arch Ration Mech Anal. 1960;6:355–70.
101. Wang CC. The principle of fading memory. Arch Ration Mech Anal. 1965;18:343–66.
102. Coleman BD, Mizel V. On the general theory of fading memory. Arch Ration Mech Anal. 1968;29:18–31.
103. Saut JC, Joseph DD. Fading memory. Arch Ration Mech Anal. 1982;81:53–95.
104. Joseph DD. Stability of fluid motions II. New York: Springer; 1976.
105. Joseph DD. Rotating simple fluids. Arch Ration Mech Anal. 1977;66:311–44.
106. Joseph DD. The free surface on a simple fluid between cylinders undergoing torsional oscillations, part I, theory. Arch Ration Mech Anal. 1976;66(4):332–43.

107. Pipkin AC, Owen DR. Nearly viscometric flows. Phys Fluids. 1967;10:836–43.
108. Zahorski S. Flows with proportional stretch history. Arch Mech Stos. 1972;24:681.
109. Zahorski S. Motions with superposed proportional stretch histories as applied to combined steady and oscillatory flows of simple fluids. Arch Mech Stos. 1973;25:575.
110. Siginer DA. On the effect of boundary vibration on Poiseuille flow of an elastico-viscous liquid. J Fluid Struct. 1992;6:719–48.
111. Siginer DA. On the pulsating pressure gradient driven flow of viscoelastic liquids. J Rheol. 1991;35(2):270–312.
112. Beavers GS. The free surface on a simple fluid between cylinders undergoing torsional oscillations, part II: experiments. Arch Ration Mech Anal. 1976;62(4):323–52.
113. Rivlin RS, Ericksen JL. Stress-deformation relations for isotropic materials. J Ration Mech Anal. 1955;4:323–425.
114. Joseph DD. Instability of the rest state of fluids of arbitrary grade greater than one. Arch Ration Mech Anal. 1981;75(3):251–6.
115. Rivlin RS. Further remarks on the stress-deformation relations for isotropic materials. J Ration Mech Anal. 1955;4:681–702.
116. Siginer DA. Memory integral constitutive equations in periodic flows and rheometry. Int J Polym Mater. 1993;21:45–56.
117. Siginer DA. Interface shapes in a torsionally oscillating layered medium of viscoelastic liquids. Acta Mech. 1987;66:233–49.
118. Siginer DA, Letelier M. Pulsating flow of viscoelastic fluids in tubes of arbitrary shape, part II: secondary flows. Int J Non Lin Mech. 2002;37(2):395–407.
119. Letelier M, Siginer DA, Caceres M. Pulsating flow of viscoelastic fluids in tubes of arbitrary shape, part I: longitudinal field. Int J Non Lin Mech. 2002;37(2):369–93.
120. Fosdick RL, Rajagopal KR. Thermodynamics and stability of fluids of third grade. Proc Roy Soc Lond A. 1980;369(1738):351–77.
121. Müller I, Wilmanski K. Extended thermodynamics of a non-Newtonian fluid. Rheol Acta. 1986;25:335–49.
122. Dunn JR, Fosdick RL. Thermodynamics, stability and boundedness of fluids of complexity 2 and fluids of second grade. Arch Ration Mech Anal. 1974;56(3):191–252.
123. Fosdick RL, Rajagopal KR. Anomalous features in the model of "second order fluids". Arch Ration Mech Anal. 1979;70(2):145–52.
124. Müller I. Thermodynamics. London: Pitman; 1985.
125. Lebon G, Cloot A. An extended thermodynamic approach to non-Newtonian fluids and related results in Marangoni instability problem. J Non-Newton Fluid Mech. 1988;28 (1):61–76.
126. Depireux N, Lebon G. An extended thermodynamics modeling of non-Fickian diffusion. J Non-Newton Fluid Mech. 2001;96(1–2):105–17.
127. Kluitenberg GA. On the thermodynamics of viscosity and plasticity. Physica. 1963;29 (6):633–52.
128. Kluitenberg GA. A unified thermodynamic theory for large deformations in elastic media and in Kelvin (Voight) media and for viscous fluid flow. Physica. 1964;30(10):1945–72.
129. Kluitenberg GA. On heat dissipation due to irreversible mechanical phenomena in continuous media. Physica. 1967;35(2):177–92.
130. Ziegler H. Some extremum principles in irreversible thermodynamics. In: Sneddon IN, Hill R, editors. Progress in solid mechanics, vol. 4. New York: North Holland; 1963.
131. Rajagopal KR, Srinivasa AR. Inelastic behavior of materials, I. Theoretical underpinnings. Int J Plast. 1998;14:945–67.
132. Rajagopal KR, Srinivasa AR. On thermomechanical restrictions of continua. Proc Roy Soc Lond A. 2004;460:631–51.
133. Onsager L. Reciprocal relations in irreversible thermodynamics, I. Phys Rev. 1931;37:405–26.

134. Prigogine I. Introduction to thermodynamics of irreversible processes. 3rd ed. New York: Interscience; 1967.
135. Karra S, Rajagopal KR. A thermodynamic framework to develop rate-type models for fluids without instantaneous elasticity. Acta Mech. 2009;205:105–19.
136. Karra S, Rajagopal KR. Development of three dimensional constitutive theories based on lower dimensional experimental data. Appl Math. 2009;54:147–76.
137. Breuer S, Onat ET. On the determination of free energy in linear viscoelastic solids. Z Angew Math Phys. 1964;15:184–91.
138. Day WA. Reversibility, recoverable work and free energy in linear viscoelasticity. Q J Mech Appl Math. 1970;23:1–15.
139. Graffi D. Sull'espressione Analitica di Alcune Grandezze Termodinamiche nei Materiali con Memoria. Rend Semin Mat Univ Padova. 1982;68:17–29.
140. Graffi D. Ancora Sull'espressione dell'energia Libera nei Materiali con Memoria. Atti Acc Scienze Torino. 1986;120:111–24.
141. Morro A, Vianello M. Minimal and maximal free energy for materials with fading memory. Boll Un Mat Ital A. 1990;4:45–55.
142. Graffi D, Fabrizio M. Non Unicita dell'energia Libera per Materiali Viscoelastici. Atti Accad Naz Lincei. 1990;83:209–14.
143. Fabrizio M, Morro A. Mathematical problems in linear viscoelasticity. Philadelphia: SIAM; 1992.
144. Fabrizio M, Giorgi C, Morro A. Free energies and dissipation properties for systems with memory. Arch Ration Mech Anal. 1994;125:341–73.
145. Fabrizio M, Giorgi C, Morro A. Internal dissipation, relaxation property and free energy in materials with fading memory. J Elasticity. 1995;40:107–22.
146. Del Piero G, Deseri L. On the analytic expression of the free energy in linear viscoelasticity. J Elasticity. 1996;43:247–78.
147. Deseri L, Gentili G, Golden JM. An explicit formula for the minimum free energy in linear viscoelasticity. J Elasticity. 1999;54:141–85.
148. Fabrizio M, Golden JM. Maximum and minimum free energies for a linear viscoelastic material. Q Appl Math. 2002;60(2):341–81.
149. Deseri L, Fabrizio M, Golden JM. The concept of a minimal state in viscoelasticity: new free energies and applications to PDEs. Arch Ration Mech Anal. 2006;181(1):43–96.
150. Amendola G. The minimum free energy for incompressible viscoelastic fluids. Math Meth Appl Sci. 2006;29:2201–23.
151. Amendola G. Free energies for incompressible viscoelastic fluids. Q Appl Math. 2010;68 (2):349–74.
152. Amendola G, Fabrizio M. Maximum recoverable work for incompressible viscoelastic fluids and application to a discrete spectrum model. Differ Integr Equat. 2007;20(4):445–66.
153. Breuer S, Onat ET. On recoverable work in linear viscoelasticity. Z Angew Math Phys. 1964;15:13–21.
154. Golden JM. Free energy in the frequency domain: the scalar case. Q Appl Math. 2000;58(1):127–50.
155. Gentili G. Maximum recoverable work, minimum free energy and state space in linear viscoelasticity. Q Appl Math. 2002;60(1):153–82.
156. Coleman BD, Owen DR. A mathematical foundation of thermodynamics. Arch Ration Mech Anal. 1974;54:1–104.
157. Fabrizio M, Lazzari B. On asymptotic stability for linear viscoelastic fluids. Differ Integr Equat. 1993;6(3):491–505.
158. Noll W. A new mathematical theory of simple materials. Arch Ration Mech Anal. 1972;48:1–50.
159. Fabrizio M, Morro A. Reversible processes in thermodynamics of continuous media. J Non-Equil Thermody. 1991;16:1–12.

160. Volterra V. Theory of functional and of integral and integro-differential equations. London: Blackie & Son Limited; 1930.
161. Stokes GG. On the theories of the internal friction of fluids in motion, and of the equilibrium and motion of elastic solids. Trans Camb Phil Soc. 1845;8:287–305.
162. Du Buat M. Principes d'Hydraulique et de Pyrodynamique. Paris: Didot; 1786.
163. Barus C. Isotherms, isopiestics and isometrics relative to viscosity. Am J Sci. 1893;45:87–96.
164. Höglund E. Influence of lubricant properties on elastohydrodynamic lubrication. Wear. 1999;232:176–84.
165. Roelands CJA. Correlation aspects of the viscosity-temperature-pressure relationship of lubricating oils. PhD dissertation, Technische Hogeschool Delft, The Netherlands; 1966.
166. Irving JB, Barlow AJ. An automatic high pressure viscometer. J Phys. 1971;E4:232–6.
167. Paluch M, Dendzik Z, Rzoska SJ. Scaling of high-pressure viscosity data in low molecular-weight glass-forming liquids. Phys Rev B. 1999;60:2979–82.
168. Bridgman PW. The physics of high pressure. New York: Macmillan; 1931.
169. Bridgman PW. The effect of pressure on the viscosity of forty three pure fluids. Proc Am Acad Arts Sci. 1926;61:57–99.
170. Andrade EC. Viscosity of liquids. Nature. 1930;125:309–10.
171. Griest EM, Webb W, Schiessler RW. Effect of pressure on viscosity of higher hydrocarbons and their mixtures. J Chem Phys. 1958;29:711–20.
172. Johnson KL, Cameron R. Shear behaviour of elastohydrodynamic oil films at high rolling contact pressures. Proc Inst Mech Eng. 1967;182:307–19.
173. Johnson KL, Tevaarwerk JL. Shear behaviour of elastohydrodynamic oil films. Proc Roy Soc Lond A. 1977;356:215–36.
174. Bair S, Winer WO. The high pressure high shear stress rheology of liquid lubricants. J Tribol. 1992;114:1–13.
175. Bair S, Kottke P. Pressure-viscosity relationships for elastohydrodynamics. Tribol Trans. 2003;46:289–95.
176. Szeri AZ. Fluid film lubrication: theory and design. Cambridge: Cambridge University Press; 1998.
177. Dowson D, Higginson GR. Elastohydrodynamic lubrication, the fundamentals of roller and gear lubrication. Oxford: Pergamon; 1966.
178. Rajagopal KR. On implicit constitutive theories for fluids. J Fluid Mech. 2006;550:243–9.
179. Spencer AJM. Theory of invariants. In: Eringen AC, editor. Continuum physics, vol. 3. New York: Academic; 1975.
180. Málek J, Necas J, Rajagopal KR. Global analysis of the flows of fluids with pressure dependent viscosities. Arch Ration Mech Anal. 2002;165:243–69.
181. Franta M, Málek J, Rajagopal KR. On steady flows of fluids with pressure and shear dependent viscosities. Proc Roy Soc Lond A. 2005;461(2055):651–70.
182. Gauss CF. Über ein neues allgemeines Grundegesetz der Mechanik. J Reine Angew Math. 1829;4:232–5.
183. Rajagopal KR. On implicit constitutive theories. Appl Math. 2003;48(4):279–319.
184. Rajagopal KR, Srinivasa AR. On the nature of constraints for continua undergoing dissipative processes. Proc Roy Soc Lond A. 2005;461:2785–95.
185. Hron J, Málek J, Rajagopal KR. Simple flows of fluids with pressure dependent viscosities. Proc Roy Soc Lond A. 2001;457:1603–22.
186. Bair S, Khonsari M, Winer WO. High pressure rheology of lubricants and limitations of the Reynolds equation. Tribol Int. 1998;31:573–86.
187. Reynolds O. On the theory of lubrication and its application to Mr Tower's experiments. Phil Trans Roy Soc Lond. 1886;177:159–209.
188. Rajagopal KR, Szeri AZ. On an inconsistency in the derivation of the equations of elastohydrodynamic lubrication. Proc Roy Soc Lond A. 2003;459:2771–86.
189. Johnson Jr MW, Segalman D. Model for viscoelastic fluid behavior which allows non-affine deformation. J Non-Newton Fluid Mech. 1977;2:255–70.

190. Gordon RJ, Schowalter WR. Anisotropic fluids theory: a different approach to the dumbbell theory of dilute polymer solutions. Trans Soc Rheol. 1972;16:79–97.
191. White JL, Metzner AB. Development of constitutive equation for polymer melts and solutions. J Appl Polymer Sci. 1963;7:1867–89.
192. Leonov AI. Non-equilibrium thermodynamics and rheology of viscoelastic polymer media. Rheol Acta. 1976;15:85–98.
193. Larson RG. Constitutive equation for polymer melts based on partially extending strand convention. J Rheol. 1984;28:545–71.
194. Bird RB, Armstrong RC, Hassager O, Curtiss CF. Dynamics of polymeric liquids. Volume 2: kinetic theory. 2nd ed. New York: Wiley; 1987.
195. Rivlin RS, Sawyers KN. Non-linear continuum mechanics of viscoelastic fluids. Annu Rev Fluid Mech. 1971;3:117–46.
196. Larson RG, Monroe K. The BKZ as an alternative to the Wagner model for fitting shear and elongational flow data of an LPDE melt. Rheol Acta. 1984;23(1):10–3.
197. Wagner MH, Raible T, Meissner J. Tensile stress overshoot in uniaxial extension of a LDPE melt. Rheol Acta. 1979;18(3):427–8.
198. Wagner MH, Demarmels A. A constitutive analysis of extensional flows of polyisobutylene. J Rheol. 1990;34(6):943–58.
199. Papanastasiou AC, Scriven LE, Macosko CW. Integral constitutive equation for mixed flows: viscoelastic characterization. J Rheol. 1983;27(4):387–410.
200. Luo XL, Tanner RI. Finite element simulation of long and short circular die extrusion experiments using integral models. Int J Numer Meth Eng. 1988;25:9–22.
201. Truesdell C, Noll W. The non-linear field theories of mechanics. In: Flugge S, editor. Handbuch der Physik. Berlin: Springer; 1965.
202. Rutkevich IM. Steady flow of a viscoelastic fluid in a channel with permeable walls. J Appl Math Mech. 1969;33:573–80.
203. Rutkevich IM. Propagation of small perturbations in a viscoelastic fluid. J Appl Math Mech. 1970;34:35–50.
204. Leonov AI. Analysis of simple constitutive equations of viscoelastic liquids. J Non-Newton Fluid Mech. 1992;42(3):323–50.
205. Kwon Y, Leonov AI. On Hadamard-type stability of single integral constitutive equations of viscoelastic liquids. J Non-Newton Fluid Mech. 1993;47:77–91.
206. Kwon Y, Leonov AI. Stability constraints in the formulation of viscoelastic constitutive equations. J Non-Newton Fluid Mech. 1995;58(1):25–46.
207. Dupret F, Marchal JM. Loss of evolution in the flow of viscoelastic fluids. J Non-Newton Fluid Mech. 1986;20:143–71.
208. Verdier C, Joseph DD. Change of type and loss of evolution of White-Metzner model. J Non-Newton Fluid Mech. 1989;31(3):325–43.
209. Kwon Y, Leonov AI. On instability of single-integral constitutive equations for viscoelastic liquids. Rheol Acta. 1994;33(5):398–404.

Chapter 3
Epilogue

A unified field theory for viscoelastic, in particular, non-linear viscoelastic fluids along the lines of the Navier–Stokes equations for Newtonian fluids is still lacking. Several constitutive structures popularly used are subject to Hadamard and dissipative instabilities as a result of the omission of some fundamental continuum mechanics principles in their derivation as well as thermodynamic principles. Naturally, they have a bearing on the ability of the constitutive formulations in predicting flows in general geometries and for high *We*. A case in point is flow in arbitrary cross-sectional tubes. The longitudinal field for both generalized Newtonian and viscoelastic fluids can be predicted quite well with existing constitutive formulations, but that is not necessarily the case for the transversal field [1].

Hadamard type of instabilities may lead to blowup in numerical computations, and dissipative instabilities may preclude the use of many popular constitutive equations at high *De* numbers prevalent in materials processing industry. However, that being said the existing plethora of equations in use does provide good predictions and sometimes even quantitative predictions at low *De* numbers in experimental laboratory settings and for dilute fluids if the use of a particular constitutive equation is restricted to a class of motions. For instance, the well-known upper convected Maxwell model is quite good in predicting elongational flows at low elongation rates, but it is totally useless in predicting secondary flows. A plethora of CEs that may be stable in the Hadamard and dissipative sense and are compliant with the principles of thermodynamics predict no second normal stress difference at all. The search for a universal CE for non-linear viscoelastic fluids has led so far down many blind alleys, and it is by no means certain that success in this respect will manifest itself in the near future.

Reference

1. Siginer DA. Dynamics of tube flow of viscoelastic fluids. New York: Springer; 2014.

D.A. Siginer, *Stability of Non-Linear Constitutive Formulations for Viscoelastic Fluids*, SpringerBriefs in Applied Sciences and Technology 14,
DOI 10.1007/978-3-319-02417-2_3, © The Author(s) 2014

ERRATUM TO

Chapter 2
Constitutive Formulations

D.A. Siginer, *Stability of Non-Linear Constitutive Formulations for Viscoelastic Fluids*, SpringerBriefs in Applied Sciences and Technology 14, DOI 10.1007/978-3-319-02417-2, pp. 9–88, © The Author(s) 2014

DOI 10.1007/978-3-319-02417-2_4

On page 72, in the caption of Figure 2.9, incorrect citation has been given; the correct citation should be as follows:

Fig. 2.9 Streamlines—flow over a slot—flow is from left to right: Constant viscosity (**a**) $\alpha = 0$; pressure dependent viscosity—Barus equation (**b**) $\alpha = 100$, (**c**) $\alpha = 200$, and (**d**) $\alpha = 360$. (Adapted from Hron et al. [185] with permission).

The online version of the original chapter can be found at http://dx.doi.org/10.1007/978-3-319-02417-2_2

Appendix

The first three Fréchet derivatives $\mathbf{S}^{(n)} n = 1,2,3$ in (2.24) evaluated at the rest state $\mathbf{G}_o$ expressed in terms of the first Rivlin–Ericksen tensor $\mathbf{A}_1$ alone [see (2.26) for the definition of Rivlin–Ericksen tensors] and new kernel functions derived from the kernel functions in (2.17)–(2.22) are collected in this Appendix [1–6].

$$\mathbf{S}^{(1)} = \int_0^\infty G(s)\mathbf{A}_1^{(1)}(s)\,ds, \quad \mathbf{A}_1^{(1)}(s) = \mathbf{A}_1\left[\mathbf{u}^{(n)}(\mathbf{X}, t-s)\right],$$

$$\mathbf{A}_1^{(2)}(t-s) = \left.\frac{\partial^2 \mathbf{A}_1[\mathbf{G}(\mathbf{X}, t-s)]}{\partial \varepsilon^2}\right|_{\varepsilon=0}$$

where $\mathbf{G}(\mathbf{X}, s)$ represents the strain history on the particle located at $\mathbf{X}$ at rest.

$$\mathbf{S}^{(2)} = \int_0^\infty G(s)\left[\mathbf{A}_1^{(2)}(s) + \mathbf{L}_1(s)\right]ds + \int_0^\infty\int_0^\infty \gamma(s_1, s_2)\mathbf{A}_1^{(1)}(s_1)\,\mathbf{A}_1^{(1)}(s_2)\,ds_1 ds_2$$

$$\begin{aligned}
\mathbf{S}^{(3)} = &\int_0^\infty G(s)\left[\mathbf{A}_1^{(3)}(s) + \mathbf{L}_2(s) + \frac{1}{3}\mathbf{L}_3(s) + \mathbf{L}_4(s)\right]ds \\
&+ \int_0^\infty\int_0^\infty \gamma(s_1, s_2)\left[\mathbf{A}_1^{(1)}(s_1)\mathbf{A}_1^{(2)}(s_2) + \mathbf{A}_1^{(2)}(s_1)\mathbf{A}_1^{(1)}(s_2) + \mathbf{A}_1^{(1)}(s_1)\mathbf{L}_1(s_2)\right. \\
&\left. + \mathbf{L}_1(s_1)\mathbf{A}_1^{(1)}(s_2)\right]ds_1 ds_2 \\
&+ \int_0^\infty\int_0^\infty 2\alpha(s_1, s_2)\left[\nabla \mathbf{U}^{(1)}(s_1) \bullet \nabla \boldsymbol{\xi}^*(s_1)\right]\ \mathbf{A}_1^{(1)}(s_2)\,ds_1 ds_2 \\
&+ \int_0^\infty\int_0^\infty\int_0^\infty \left\{\sigma_1(s_1, s_2, s_3)\mathbf{A}_1^{(1)}(s_1)\mathbf{A}_1^{(1)}(s_2) + \sigma_4(s_1, s_2, s_3)\,tr\left[\mathbf{A}_1^{(1)}(s_1)\mathbf{A}_1^{(1)}(s_2)\right]\right\} \\
&\mathbf{A}_1^{(1)}(s_3)\,ds_1 ds_2 ds_3
\end{aligned}$$

D.A. Siginer, *Stability of Non-Linear Constitutive Formulations for Viscoelastic Fluids*, SpringerBriefs in Applied Sciences and Technology 14,
DOI 10.1007/978-3-319-02417-2, © The Author(s) 2014

$$\mathbf{L}_j = \boldsymbol{\xi}^* \bullet \nabla \mathbf{A}_1^{(j)} + \mathbf{A}_1^{(j)} \bullet \nabla \boldsymbol{\xi}^* + \left(\mathbf{A}_1^{(j)} \bullet \nabla \boldsymbol{\xi}^*\right)^{\mathrm{T}}, \quad j = 1, 2$$

$$\mathbf{L}_3 = \boldsymbol{\xi}^* \bullet \nabla \mathbf{L}_1 + \mathbf{L}_1 \bullet \nabla \boldsymbol{\xi}^* + (\mathbf{L}_1 \bullet \nabla \boldsymbol{\xi}^*)^{\mathrm{T}}$$

$$\mathbf{L}_4 = \mathbf{L}_\xi \bullet \nabla \mathbf{A}_1^{(1)} + \mathbf{A}_1^{(1)} \bullet \nabla \mathbf{L}_\xi + \left(\mathbf{A}_1^{(1)} \bullet \nabla \mathbf{L}_\xi\right)^{\mathrm{T}}$$

$$\mathbf{L}_\xi = \left(\boldsymbol{\xi}^{**} - \left.\frac{d\boldsymbol{\xi}}{d\varepsilon}\right|_{\varepsilon=0} \bullet \nabla \boldsymbol{\xi}^* - \frac{1}{2}\boldsymbol{\xi}^*\right) \bullet \nabla \boldsymbol{\xi}^*$$

$$\boldsymbol{\xi}^* = \int_t^\tau \mathbf{u}^{(1)}\left(\mathbf{X}, \tau'\right) d\tau', \quad \boldsymbol{\xi}^{**} = \int_t^\tau \mathbf{u}^{(2)}\left(\mathbf{X}, \tau'\right) d\tau', \quad t > \tau$$

$$\mathbf{u}(\mathbf{X}, t; \varepsilon) = \varepsilon^n \mathbf{u}^{(n)}(\mathbf{X}, t)$$

The shear modulus $G(s)$ and the quadratic shear modulus $\gamma(s_1, s_2)$ together with the kernel functions $\alpha(s_1, s_2)$, $\sigma_1(s_1, s_2, s_3)$, $\sigma_4(s_1, s_2, s_3)$ are derived from and related to the kernel functions $\zeta(s)$, $\beta_{21}(s_1, s_2, s_3)$, $\beta_{22}(s_1, s_2, s_3)$, $\beta_{31}(s_1, s_2, s_3)$, $\beta_{34}(s_1, s_2, s_3)$ in (2.19)–(2.22)

$$\zeta = \frac{dG}{ds}, \quad \beta_{21} = \frac{\partial^2 \gamma}{\partial s_1 \partial s_2}, \quad \beta_{22} = \frac{\partial^2 \alpha}{\partial s_1 \partial s_2}, \quad \beta_{31} = \frac{\partial^3 \sigma_1}{\partial s_1 \partial s_2 \partial s_3}, \quad \beta_{34} = \frac{\partial^3 \sigma_4}{\partial s_1 \partial s_2 \partial s_3}$$

References

1. Siginer DA. On the effect of boundary vibration on Poiseuille flow of an elastico-viscous liquid. J Fluid Struct. 1992;6:719–48.
2. Siginer DA. On the pulsating pressure gradient driven flow of viscoelastic liquids. J Rheol. 1991;35(2):270–312.
3. Siginer DA. Memory integral constitutive equations in periodic flows and rheometry. Int J Polym Mater. 1993;21:45–56.
4. Siginer DA. On some nearly viscometric flows of viscoelastic liquids. Rheol Acta. 1991;30(2):159–75.
5. Siginer DA. Oscillating flow of a simple fluid in a pipe. Int J Eng Sci. 1991;29(12):1557–67.
6. Siginer DA, Valenzuela-Rendón A. Energy considerations in the flow enhancement of visco-elastic liquids. J Appl Mech. 1993;60(2):344–52.